T0220035

SpringerBriefs in Ethics

More information about this series at http://www.springer.com/series/10184

Kathleen Benton

The Skill of End-of-Life Communication for Clinicians

Getting to the Root of the Ethical Dilemma

 Springer

Kathleen Benton
Clinical Ethics and Palliative Care
Armstrong State University
Savannah, GA, USA

Quotation by Angelo E. Volandes
Copyright © Angelo E. Volandes, 2015
The Conversation: A Revolutionary Plan for End-of-Life Care, Bloomsbury Publishing Inc.

Epilogue quotation by Elisabeth Kübler-Ross and David Kessler
Copyright © Elisabeth Kübler-Ross and David Kessler, 2007
On Grief and Grieving: Finding the Meaning of Grief through the Five Stages of Loss

ISSN 2211-8101   ISSN 2211-811X   (electronic)
SpringerBriefs in Ethics
ISBN 978-3-319-60443-5   ISBN 978-3-319-60444-2   (eBook)
DOI 10.1007/978-3-319-60444-2

Library of Congress Control Number: 2017943982

Printed on acid-free paper

This Springer imprint is published by Springer Nature
The registered company is Springer International Publishing AG
The registered company address is: Gewerbestrasse 11, 6330 Cham, Switzerland

To Daniel, who strongly advised every clinician he befriended or talked with to put on a face for your patients, to "fake it until you make it," if necessary, because patients are the vulnerable and despaired. He encouraged them to bend the rules when it resulted in patient dignity and mercy, to talk with peers so the right hand knows what the left hand is doing, and to know your patient's name because—even though you have multiple charts—the patient gets only the one name, the one life.

And to the family who supported Daniel and me through our lifelong drama. Daniel and I are like-souls and those who put up with us are saints.

Foreword by Kevin T. FitzGerald

In the *Nicomachean Ethics* (Book 1, Chap. 3), Aristotle recognizes that each person judges best the issues with which the person is most familiar. If we who are involved in clinical ethics accept this premise, then we need to recognize the special experience and expertise Kathleen DeLoach Benton brings to the many challenges and opportunities present at the end of life. From her years of sharing her brother's struggle with Proteus syndrome, and her years working in clinical ethics, the author fashions an approach to caring for a dying patient—and that patient's loved ones—that directly addresses the lack of ability and skills regarding end-of-life communication too often found among healthcare professionals. Though we are focused on pursuing healing and health for all our patients, we often falter in this pursuit when faced with a dying patient—perhaps because we feel at a loss to be able to provide any healing in such situations.

In this book, the author confronts this feeling of helplessness and our consequent desire to avoid such situations and reminds us of the healing power of true compassion ("suffering with") and honest communication. Out of her own experience as a loving sister and a caring professional, the author describes how the entire healthcare team can work together to ease the struggles of the patient, family, and friends by taking more of the communication burden on themselves as a team—a very real challenge in the high-paced, high-pressure environment of a hospital. In the complex, and often messy, realities of dying, with the patient surrounded by machines, it is imperative that the healthcare team prevent the patient and loved ones from being battered by information and confused by contrasting messages and communication styles. To achieve the quality of care we all aspire to give, the healthcare team must coordinate well to provide the time, attention, and empathy required to prepare the patient and loved ones as best as possible for the dying process with all its knowns and unknowns.

Kathleen DeLoach Benton has written a challenging book for a challenging issue. You may not agree with the author on every aspect of her approach or in every case, but to engage substantively and sincerely the issue of communicating well with the dying patient and his or her loved ones in your own practice is to accomplish what this books intends.

Dr. David P. Lauler Chair in Catholic   Kevin T. FitzGerald
   Healthcare Ethics
Pellegrino Center for Clinical Bioethics
Georgetown University Medical Center
Washington, DC, USA

Foreword by Ira Byock

If you presume the worst thing that could happen is death, doing everything possible to keep someone alive makes sense. But many of us have seen patients, friends, or relatives die badly, and we recognize that there are states worse than death. Every treatment with potential benefits also brings known burdens and risks. Since illness is fundamentally personal and no two persons are exactly alike, a treatment plan that represents the best care for one person might be utterly wrong for another. That's why, in healthcare today, conversations that enable clinicians, patients, and families to match medical treatments to people's values and personal priorities are essential to quality. How people wish to be cared for when recovery is unlikely and life is waning must become an integral part of those conversations.

Providence Institute for Human Caring   Ira Byock
Torrance, CA, USA
Dartmouth's Geisel School of Medicine
Hanover, NH, USA

Foreword by Renzo Pegoraro

Communication is an essential requirement to face together—physicians and nurses, patients, and families—the end-of-life care and the ethical issues involved.

The personal existential experience and the work of clinical ethics consultation enable this book to confirm the need for a good and effective communication at the end of life, stimulating education and training to develop communication skills.

It is important to recognize the different illness processes, diversity of age and contest, and difference of religious and cultural sensibility, as the author presents through many and useful clinical cases. But in all these experiences, strong communication offers the possibility for a real encounter to mend fears, suffering, misunderstanding, or mistakes. In this perspective, physicians and nurses can realize good care to prevent loneliness, to relieve pain and suffering, to avoid risks of neglect, or to request euthanasia.

Pontifical Academy for Life   Renzo Pegoraro
Vatican City, UK

Foreword by Christian Sinclair

When it comes to difficult conversations about serious illness, there is room for improvement for clinicians of all disciplines. Collectively, clinicians contribute to the shared understanding patients and families have about their illness experience. The ability to help patients make challenging decisions in the face of difficult odds may hinge on how the physical therapist or night nurse communicates about care. Any clinician may change the eventual clinical outcomes through even apparently trivial acts such as repeating clichés ("Gotta take it day by day, and keep fighting.") or reducing anxiety ("I know the palliative care team. They've been so valuable in difficult times.") Yet, by placing a significant training focus on the biomedical model, we devalue the most critical treatment in our healthcare toolbox, communication. The Institute of Medicine 2014 report on *Dying in America: Improving Quality and Honoring Individual Preferences Near the End of Life* focused on building communication skills, not only in palliative care specialists but across specialties and disciplines. This book by Kathleen Benton helps take the IOM goals and places them within reach for frontline clinicians.

American Academy of Hospice and Palliative Medicine   Christian Sinclair
Chicago, IL, USA
Pallimed, Alexandria, VA, USA

Preface

Although I've tried to keep the medical jargon to a minimum so that this book can be available to all readers, *The Skill of End-of-Life Communication: Getting to the Root of the Ethical Dilemma* is meant for those in medical schools and advanced allied professional schools. As I outline in Chap. 1, what my family and I—and most importantly, my brother Daniel—went through during his long illness and particularly his final years has shaped our outlook of how we as medical personnel advise, treat, and very unfortunately dismiss persons with terminal illnesses.

Death is inevitable, whether we meet it in a hospital, at home, on the streets, or in the wilderness. It is the great unknown. Bound by the cultural influences of medicine, medical professionals feel much more at ease doing more for our patients rather than less. Family members, however, are left to ride an emotional roller-coaster controlled by whichever physician or other professional walks in the room, beckons them to a meeting, or imposes an opinion that is mostly dependent on the spectrum of culture he or she hails from. There are doctors, nurses, and healthcare staff who fear death and who impose a sense of failure or loss on the patient and family. There are those who ignore the specter of death, who go implacably from one assigned duty or one predetermined test to another in a rote march toward the end of their ward round. There are also providers who embrace death and try to shift our thinking for a minute. But that minute is grueling; it is painful, not welcomed by many these days. As such, that minute becomes smothered underneath false hope and is buried beneath the discomfort of communication and ease of comfort found in routine and standards of care.

Some patients will be eased into death, and some will fight it as if it is the lifelong battle they have always prepared for. My brother, among the latter group, was not an enigma or an anomaly; he was a certain type; there are others. I've learned that you do not ignore these types. You can walk with your patients through the end of life if you agree to meet them where they are. If not, you are abandoning them to go at it alone with newer, stranger providers. You have to choose, and it is rarely an easy choice.

It is so hard to know when to stop the artificial support. In theory, our modern attempts to prolong life mostly seem excessive. It seems unattractive to go on life support, yet when fear supersedes ideology, many ill people walk around wanting us to "do everything, resuscitate and all" until death is imminent. We consider death in a hospital too sterile and uncomfortable, yet the vast majority of us die there anyway.

For those patients who choose to die at home, if wishes aren't sought before the body becomes reliant on machines, a land mine of disconnect occurs when a provider is ignorant of the logistics of acceptable versus unrealistic discharge plans. When a member of the care team fails to learn the whole story of a patient's care, he or she is not an informed provider. If patients must be informed, so too should the members of the treatment team. In my brother Daniel's case, the doctors had created—unintentionally but quite officially—a discharge planning nightmare. A conscientious provider knows that if you recommend any artificial device without educating yourself on what happens after discharge, you've done patients and their caregivers a disservice. Rules and regulations for what nursing homes accept and what families can provide are different, but the long and short of it is that you have to have someone with medical skills, no need for sleep, an accepting payer source, never-ending patience, and at least a rudimentary understanding of patient psychology, as well as a thousand new items in the patient's habited environment, possibly including such diverse items as industrial medical shelving to take care of a patient on peritoneal dialysis, a trach, and a vent, not to mention all of the tools that go with such necessities. One of the benefits and detriments of machines is that they do not go to bed, nor do they need bathroom breaks, food, and other such necessities. But they break, malfunction, and require full power, day and night, during storms and power outages.

What alternative environments do patients have, once they have entered the land of artificial support? Daniel could not go into hospice because he required too much technology; at the same time, he presented as too much liability and needed too much skilled time for home health assistance; for home aides, Daniel needed too much skill in general; and for family and friends, as good-willing and good-hearted as they were, their training was limited and they likely had obligations of their own to consider.

So what happens when communication is missed, when all this becomes messy? I know it is a comfort to the provider to distance himself or herself from the patient and discuss the failing organ—whether liver, heart, lungs, or kidney—in practical, unemotional terms. But is this what a patient needs? In an effort to give the patient every chance at life, we do everything we can to prolong existing until we have no more viable options—and then frustrations lead to bad news delivered poorly.

As you read Chap. 1 and consider Daniel's progression, notice the increase in support and his willingness to accept a new quality of life in exchange for continued life. It is important to tell this story because within the anecdotes of Daniel's life and end of life, the reader will discover all the communication deficits and the ethical conundrums that have been born out of artificially supporting life without first a conversation on what that life will be like.

Although it is necessary and appropriate to have a highly specialized palliative clinical team involved in end-of-life care decisions, these conversations are too common and too necessary to rely only on trained specialists. So this book is written for all the clinicians, the nurses, the nutritionists, the respiratory and physical therapists, and all the caregivers who are traditionally excused from participating in these dialogues but whose support and input are essential. And it is written for physicians, who are expected to weigh in on these dialogues but rarely do so in a substantial and helpful way.

My intention of writing for medical and allied professionals is to breach the walls that our healthcare silos create. Silos are inherent to medical care. The disciplines work together yet are miles apart. The social worker is not speaking to the physician, and the nurse is not speaking to the respiratory team. The goal of charting, rounds, and electronic medical records has begun to close these gaps. However, the tenuous and demanding nature of healthcare, along with the inherent hierarchy that admittedly (or un-admittedly) exists, inhibits a crossover of clear communication. And thus, a patient and his or her family are faced with the unorthodox challenge of understanding a message from multiple professionals with multiple spectrums and spans of knowledge and with varying opinions on how far treatment should proceed. The continuum of care is broken, gapped, and challenged, particularly when the ability to heal is no longer likely and when the care that can be achieved is only one of compassion and comfort. Thus, it is more important than ever that this communication be a team approach.

It would be a mistake to only write this book for the doctor. All it takes is one allied professional who feels the need to give inaccurate or false hope to throw the family into confusion and second guessing. When a member of the care team is communicating clearly that the end is near, that individual can quickly become the bad guy if other more hopeful opinions challenge a terminal likelihood and generates discontent for a family struggling with grief. It is left to the nurse who is with the patient 12 hour a day, the respiratory therapist who earns the trust for breath, the physical therapist called in to sometimes rehabilitate those who cannot be rehabilitated, and the social worker involved to identify all family issues; together, not individually, they can develop a discharge plan, even when there are few accepting facilities. Thus, the team must all be delivering the same prognostic message. Poor prognosis is one of the most difficult types of information to convey. Repetition becomes the key to processing and acceptance for both patient and family

So how do you talk someone into dying? The short answer is you don't. You meet them where they are and sit with them. Some will die in denial, some with clear focus on the path ahead. All should die in hope of something better, something not scary, less suffering, and more life. For all we know about life, living, and eventually dying, what comes next is the biggest mystery of all.

Savannah, GA, USA   Kathleen Benton

Acknowledgments

I would like to acknowledge Chloe Long for the time and expertise she gave sifting through data and research to create a well-thought-out and skillful manuscript.

I would like to acknowledge Carol Gerrin for her brilliance in biostatistics and patient data to make sense of the numbers that in the end proved trending issues all pointing to poor communication.

I would like to acknowledge Dr. Anthony Costrini for coining 'preemptive ethics' for which we embrace in the art of proactive communication.

I would, lastly, like to acknowledge Ann Beardsley, a new friend whose wisdom and clarity gave better flow and lasting understanding to the purpose of this text.

Contents

Introduction to Daniel

1

Abstract

Daniel, a 30-year old man with Proteus syndrome (elephant man disease), is introduced. After 110 surgeries, Daniel's family elects to care for him at home in the final stages of life. Problems ensue when poor communication from hospital staff and the lack of a cohesive discharge plan give neither Daniel nor his family any idea of what to expect upon leaving the care of the hospital. In this chapter, Daniel, the author's brother, is used to demonstrate the complexities of end-of-life issues, even for families who are familiar with the medical system.

Keywords

End-of-life conversation • End-of-life care • Advance care plan

"Kathleen, I feel like shit."

I sit in Daniel's room with him and all his anger. The anger takes up most of the space. The emotional state, his medication tolerance, and endured stamina fuel Daniel's consciousness. He's alert, oriented, and wondering what's next. I know from my years in the medical field that it's an established progression from a patient who might get better one day to a patient who needs to prepare to die, to recognize the artificial mechanism necessary to support him: first hemodialysis, then peritoneal dialysis, then feeding tube, then the vent, then pressors, then…an inevitable slide toward death. Bound by the trach in his neck, Daniel must whisper and mouth his frustrations and his demands. He is held hostage as a patient, at the mercy of every beep, blip, plug, and air bubble in the line. He is at the mercy of everything and everyone in that room who must do for him what he can no longer do for himself. Even to get a piece of ice, Daniel has to negotiate with the speech therapists,

© The Author(s) 2017

K. Benton, *The Skill of End-of-Life Communication for Clinicians*,
SpringerBriefs in Ethics, DOI 10.1007/978-3-319-60444-2_1

who fear he will choke and that they will be held liable. Swallowing—even a tiny ice chip—is difficult for Daniel because if he is not careful, he will aspirate.

We are held hostage as a family as well, because Daniel is my younger brother. We—family and patient, mom, dad, brother Michael, myself, and friends who want to talk to Daniel again—are bound by the evasive and invasive technologies that support life and sustain this near death that seems to go on forever. No matter what impending discussion we have planned with each other or with other medical professionals, the next round of dialysis, lab results, or bath interrupts.

Daniel was born with undiagnosed Proteus syndrome, more commonly aligned with the elephant man disease. He had fifteen surgeries in his first year of life, most of which were away from our hometown. By the time of his end of life, he had had 110 surgeries. It is important to understand that that my parents, my whole family (I, the eldest, and Michael, the middle child), have spent our lives wrapped up in next surgeries and hospitalizations. It did not bog us down but certainly defined us and who we would become. My parents became parents who had to fight for Daniel's life with nearly every clinical person involved in Daniel's care, from the first neonatologist who told my mother that Daniel would never live past a year, to the spinal cord surgeon who said he wouldn't walk. But he did walk and he did live, well and quite normally for many years. He went to primary and secondary school and even to the university on scholarship. He swam on the swim team, was the ball boy for his baseball team, drove, got in more trouble than I, and enjoyed life, more than most. He had always been absolutely positive and willing to endure every new treatment or surgery but never wanted to be tied down by his illness. He knew he would never be anything but normal in the ways that mattered, despite the stares and humiliation other people sent our way. He was determined not to be bound by his deformities, tumors, or orthopedic pain. He knew he would never tolerate artificial support to keep him going. His focus was status quo for a normal young man: a girlfriend, a good job, and more energy.

In his twenty-fourth year, he had a hemangioma erupt on a kidney that led to an abrupt acute hospitalization on life support and in ICU, but followed by recovery, more doctors' visits, and more concern for the future, but absolute "Daniel normalcy" otherwise. And so his first compromise occurred: He took more pain medication, witnessed abnormal renal lab results, worried he might need a transplant, and brushed with the mortality that had always been hovering.

In his twenty-seventh year, he hit the pulmonary toilet. He was told death was in sight by an abrupt specialist, and he was left wondering why the specialist was so adamant. Daniel had never had any breathing issues, but he knew the growths that sat on his spine and in his lungs were likely to eventually cause some damage. And they did—expected by those who looked at his encyclopedic-sized chart for the first time, unexpected by we who watched him live "normally." Compromise #2 occurred when he agreed to around-the-clock O_2 support, eventual BiPap (bilevel positive airway pressure to ease his breathing) at night, and palliative furosemide (Lasix, a diuretic to release extra fluids from the body but certain to kill off that last kidney). But he still socialized with friends and even traveled to California to visit our brother Michael during his military duty there.

In his twenty-eighth year, he added end-stage renal disease to his list of health concerns. And so he sat in his chair most days when he was not at dialysis, yet he still visited with friends, ate his ever-loved gourmet food, and we went to Rhode Island for another visit with our military sibling.

In his twenty-ninth year, he reached his imminent end of life, and this is how it went:

December 10—Daniel was admitted to the hospital for peritonitis. Breathing worse, he was given IV antibiotics and he managed to come home on December 23—not truly stable enough to be discharged but able to have Christmas where he felt most comfortable. He stayed in his chair at home and we brought Christmas to him. It was a familiar routine, as we had done this often when he felt bad. The party ended on January 4 when the reality of how sick he had gotten finally set in and he had to be ambulated to the hospital. It was the first time he had ridden in an ambulance, ever. Thousands of hospitalizations and he had always been well enough to go by car, but not this time. His zest for life surfaced, however, and Daniel's "cool experience" with EMS became a blast of a ride with great people to dialogue with in the one field of medicine he had never been privy too.

He started that admission in an intermediate care unit and on the BiPap pretty continuously. January 14 was his birthday. Since he couldn't see his nephews or niece (godchild), he was moved to the floor for a hospital party. During this time his go-to physician, a surgeon, had been theorizing about the possibility of an extensive surgery to restructure his thorax, since it was this structure that was crushing his lungs and thereby denying their ability to balloon as they should. This idea was quickly shot down as his respiratory state dwindled. The discussion on "trach for now" happened so quickly it felt like a slap in the face. Though it was never overtly stated, "now" might mean forever, barring a medical miracle, but nobody said that out loud. And yet, Daniel made it clear to us he was not ready and did not want to die. "If I don't do this, then what else is there?"

For Daniel, luck began to run dry. An ear, nose, throat (ENT) surgeon was brought in to place the trach and followed up exactly one week postoperatively; he did a good job, but we never saw him again. A nutrition tube (PEG) was suggested by another specialist for obvious reasons, even though Daniel had far too much scar tissue for this to be a success. Dialysis would be a necessity, and it would have to be at home, since there was no likely way he could safely transport to dialysis 3 days per week; a peritoneal dialysis site would therefore be required.

While the family, including Daniel, considered all these ramifications, on the other side of the same facility operating in the siloed disciplines normal for healthcare, an ethics consult convened to question the ordered procedures that would, as a matter of opinion, cause undue harm. They determined that taking care of Daniel would require skill, time, and 24/7 attention. Instead of calling the physicians recommending these procedures, they called the family. We wondered if this new sequence of events would lead to a quality of life Daniel did not want, and wondered which way Daniel might have the life he still wanted. But for him, anything was better than no life. Daniel was stuck on a BiPap, suffocating and desperate to live. It was almost as if there was no time or place for discussion. He consented to the recommended treatments as the lesser of two evils. The surgery occurred and Daniel morphed to a man dependent on technology to live.

Then came the vent and ICU admission "for a few days," which became a lot longer than anyone expected: eighty-plus days. There was no likely discharge once the vent got involved since Daniel could neither come off the vent nor could he get better. Four important conversations followed. The first took place with the physician in charge of intensive care, an intensivist. Standard to ICU care, this specialist takes over once a patient is admitted to ICU. The intensivist was new to the case, as were all the physicians—none of the original admitting doctors could stay on the case, per standard protocol, though some followed as a compassionate courtesy. The intensivist told the family (minus Daniel) that Daniel would likely not live past 3–6 months and that the use of life-support machines would be futile. Someone on the medical staff was supposed to relay some of this to Daniel but no one knew who that someone was. No change in care came in the days that followed, though there was some focus on whether a nasal gastric tube could be placed to improve Daniel's nutrition intake. Daniel seemed focused on this and demanded nutrition, completely oblivious to his predicted demise.

Conversation #2 included Daniel, whose biggest question was "What do you mean, end stage?" This dialogue included a new intensivist, changed every 10 days as standard procedure in ICU, as well as a specialist from the Ethics Committee, members of the support staff, and family. None of his usual and known doctors were present and this created major trust gaps since no rapport had been established for Daniel with this new ward round. This meeting was brutal, but at least Daniel was given pain medication and Valium to take some of the pressure off him; the rest of us were not so fortunate. We heard again how there was no plan for his care and that they would move him (and the vent) to an intermediate care floor—despite the usual protocol for patients in his state to be in an ICU environment. One of the staff told us, "We are moving him into intermediate, though he is critical." Staff turnover in a unit like this changes often, and it becomes difficult for even the most devoted professionals to keep up with what is going on in a patient who is not expected to heal. When Daniel transitioned to a tiny room packed full of his machines, it was clear the compassion for a young man in his twenties had degenerated. Daniel had become the "patient who is back, and what a sad case?!" The unspoken sentiment seemed to be "If you can't get well, you need to get out and give the space to someone who can." And they were right: Daniel was a man in a bed for whom thousands of dollars' worth of resources poured in and he was going nowhere. Logic tells any clinician that this long in a hospital is good for no one. Nothing about taking care of a patient like Daniel is going to feel rewarding if you don't know the person, and most staff members do not have the luxury of time. In Daniel's case, our only "friends" were the environmental staff—a few empathic nurses and the dialysis nurse whose compassion was linked with having followed Daniel for so long.. In essence, Daniel's team had no captain. Each new ward round brought a new nurse, respiratory therapist, and hospitalist physician to manage the case, but each of them had no clue what the last had offered in the way of guidance or dialogue. One asked that calories be counted; the next one stated, "You can't eat"; and a third said, "You can eat, but don't tell me because I won't be held liable." For Daniel, lost in a fog of pain meds, the only points of continuity were family and friends. His mom/dad/aunt/sister/friends made up those who received and relayed information every day from the assortment of specialists, but no one really knew the goals of care or the plan to go home.

Just thinking about moving Daniel into home care raised hundreds of questions: Who handles the issue of coding and distress? Who will manage the patient when he can't ambulate to the doctor? Who will suction his lungs? Who will handle the mucus plugs? Who is there to call when death is imminent if you do not have hospice and the hospital has nothing to offer but a defibrillator? Hospice might seem the best choice but asking a suffocating Daniel to take off the vent is not a question a friend should have to ask.

In order to take Daniel home, we had to have a boundary-setting discharge plan. Daniel would need to sign a statement of liability and a DNR order. We wondered who could talk to Daniel about death. I am and will forever be disheartened and disappointed by the medical community whenever I remember one tearful discussion with our mom. She asked, "Maybe it is our role as parents? We brought them in, must we help them leave? The doctors aren't doing it."

The palliative team was brought in but they were limited by Daniel's autonomous wishes and (though it pains me to admit it, even though I understand his feelings) the arrogance Daniel expressed by chalking them up to "palliative doctors who [did] not know him and his forever fight with, and will for life."

Talking to Daniel about death was going to be very difficult. Our mom gave it a try but couldn't continue when Daniel said, "Mom, you are trying to make me die." On my first attempt to discuss his future—or lack of it—he stated quite firmly that he knew my thoughts on prolonged suffering and he begged me not to kill him. We were all between the proverbial rock and a hard place, and were standing on quicksand. I made a choice to step in because discharge and conversations were going nowhere. I could no longer watch Daniel live in a hospital just to die in a hospital, right alongside the powerful, positive spirits that were once his and my parents.

So our third conversation was between me and Daniel to set boundaries—we changed his code status to DNR, we wanted no more hospital stays, and we invented a waiver of liability for the aide company so that they could come in and not be sued if anything happened on their watch. I pulled favors from angelic physicians in pulmonology and palliative care for home care only, and suddenly we had piecemealed a care plan and were ready to take Daniel home. Take note, this was not a legitimate care plan; our familiarity with Daniel's history and capabilities, as well as our knowledge of the health care system, enabled us to create this facilitated option because of the out-of-the-box scenario we found ourselves in: Daniel was too sick to stay home yet not well enough to leave the hospital. He was not expected to live, yet his death might not be imminent. He was healthy—except for a long, interconnected list of sicknesses and weaknesses.

The DNR order was the hardest. After multiple palliative discussions, the trust between siblings seemed to play on my brother's understanding. Since I knew he understood what I do as a professional, I tread lightly with a boundary-setting discussion. Of all the consults, this conversation made me the most nervous, and I was quick to tell him this. Other questions soon surfaced from Daniel. "Why don't they want me to fight?" "Why won't the trach fix the issue?" "Why do I have to agree to DNR, to get hospice involved?" "I won't give up dialysis or the vent … but I do want hospice." Something in him must have known that he needed good end-of-life care as much as he wanted technology, just to manage the symptoms, if for nothing else. He tearfully and angrily agreed to do the DNR and the "no more hospitalizations," though he remained unconvinced that hospitals were not a best place for him. The logistics of his needs were hard to explain. He could not wrap his mind around his dependence. "Why can't I go to my doctors when I get stronger?" "Why do I need an ambulance to go home?"

I had to be blunt with him. Anything less than that was not getting through. "Daniel, when was the last time you sat in a chair?" No answer. "That's why. That's why you are dwindling and limited. That's why this is hard. This is not fighting, Daniel. It is bullshit. No one should reside in this closet turned room for five minutes, not to mention eighty days…. Nothing good can come of staying here."

Conversation #4 happened with all the players for home care present at the hospital: the volunteer pulmonologist, hospice physician, pain management physician, palliative physician (since we didn't yet know which would manage the pain), the head of case management, case worker, parents, me, Ethics personnel, and home care staff. The hospitalist had taken over for the attending physician, but I wasn't even sure if he knew Daniel's name. It literally takes a village—in terms of people as well as equipment, once you add in all the technology and such. The palliative physician started by telling us, "He says he wants to live long enough for his life to mean something. Until that time comes, this is where we are." There were seven agencies to care for him; my parents each had the weekend, a weekday, and two nights. We thought Daniel's time with us was short, but no one really knows when a person will die. The human spirit and strong will can fool even the most experienced of hospice physicians.

At home, the house became a revolving door of caregivers and emotions. On Day 7 at home, Daniel said, "I don't know how much longer I can do this." On Day 9, he said the same thing. But on Day 10, he said, "Mom, you know I'm dying." She cried hard even though she tried not to. It feels less real when someone stays in denial, as if—even though you know their truth to be wrong—there is some part of you that allows the denial to carry you, just a little. Then, when all the family knows, when there is no more pretense, death becomes present during every conversation, every meal, every thought.

When we first talked about Daniel going home on all this artificial support, I warned my parents multiple times about the perils of denial. It is a common lay misconception, and even a professional one, that once the patient is stabilized on the technology, though quality of life is sacrificed, life is not dwindling. In actuality, the truth is that a machine is volatile, tricky, power dependent, and not malfunction resistant. The likely cause of death might be this very technology supporting life. When something happens, the family gets blamed. The vent might grow mold, the cycler might not drain, the O_2 might shut off. This is the definition of burdensome care, care that burdens the family *and* the patient. Such care is available and reasonable for some, and that is ok—as long as all parties realize what they are getting into.

A first home end-of-life conversation was necessitated when Daniel's leg hurt so badly he feared a clot and wanted to go in hospital for an ultrasound. "And then what?" I asked. "Will they give you blood thinners that lower your already low hemoglobin and will you ever get out of hospital?" He was reluctant to hear me. The leg pain eventually ceased and more issues came and went. He was dwindling and then bouncing back; his blood pressure was 90/40 (his normal blood pressure is 110/70), and infection and dialysis issues were common. He explained, "I will never give up and if I don't go to the hospital that is giving up."

I replied, "No, it's not. They simply have no ability to do anything more to help you," I answered him. "You need to recognize the limitations of medicine, and with nothing left to give you, the hospital will wither your spirit."

"I want to beat the odds; I have more to do, more goals."

"Like what?" I asked.

"Just lots of ideas," but he was just too tired to talk about or implement them.

So I rebutted, "Sometimes goals are just there to give us drive and focus, and sometimes someone initiates the goal as their brainchild and then has to pass it on to be finished."

He didn't buy into much of my mumbo-jumbo and later told my mother, "Kathleen was over with one of her end-of-life talks. I am getting sick of them." I wanted to heal his heart, give him closure, and remove the struggle at the end. He wanted to live for all the life that was left. He was right and I was wrong. He was doing it his way.

Three months later, Daniel was still hanging on. His lesson became one of demonstrating the power of the spirit and a stubborn will. Be careful not to ignore the sheer will that extends life. When scientists make the mistake of knowing that the breakdown of cells controls the end, we ignore the art involved in living. There is an art to life, and there is a similar art to end of life. Do not underestimate this art called spirit. We've all heard the stories and the anecdotes: "Grandpa waited for John to fly in to pass away." "Mom waited for us all to leave the room to take her last breath." Such a spirit, such determination, has some control, and so did Daniel. It did not matter that his O_2 with vent sat at 83–90, or his carbon dioxide was frequently high. It did not make a difference that his creatinine was 6.1 and an infection reared its ugly head once per week. He was almost admitted into the hospital over and over again. He lived and stayed alive because he wanted to. He admitted his suffering, even verbalized the idea of quitting the technology for one minute. But as fast as the thought came, it went and he lived on, bound and determined not to leave until he was ready. And for Daniel, "ready" only came in those last few minutes.

In the fourth month after discharge, the community joined together to build Daniel a room of his own, since he and all his equipment were packed into a tiny living room. This new room had windows and space, with room to roll around in and look outside. This was the best thing that could have happened to Daniel at this point, since his spinal cord tumor had creeped back and he was now looking at paralysis, which led to more compromises, some of which he never learned to accept. In his room, we placed a chair that folded to a bed, and he looked at it and said, "That will be a good place for Mom to be next to me when I die."

What a journey! What a story! And at times, what a nightmare! Grief and loss are difficult enough without the obstacles that plague this time. My profession, my experience, and Daniel gave me some insight into loss, but I am not alone in seeing the need for more improvement in communication about this time of life. In the published brief *Dying in America* (2014), the Institute of Medicine made a recommendation for better professional education (specifically, the line addresses "deficits in equipping physicians with sufficient communication skills"). Kudos to the palliative care specialty with all they have brought to the conversational table, but it should not be left up to a single discipline with a shortage of providers to make believers of those who do not see the worth of the end-of-life conversation and to further educate those who align the palliative specialty with hospice. Other experts, specialists, and clinicians need to know how to talk to dying patients and their families and caregivers, and they must be held accountable so that they avoid abandonment or deferment when conversations become uncomfortable.

It may not shift the choices patients make. An informed choice, whatever the choice, is a success. Two reasonable people may disagree on which way an end of life should unfold; the dilemma occurs when they do not get the chance to talk about it, or to *choose* not to. I want to be clear in defining the ethical dilemma. In such a conversation, there is no place for futility or injustice, and both parties should be acting in the patient's best interest or at least non-maleficence. The ethical dilemma lies in the communication gap. There is no judgment placed on a patient's decision, one like Daniel who wishes to squeeze out every ounce of quantity and compromise the quality. Likewise, there is no judgment or immorality placed on one who wishes to choose the present quality over the longevity. Neither is wrong, or against most faith or religious belief. The concern is the unknowing, all that is left unsaid when a patient or family chooses; this is the dilemma that must shift.

This book has two threads, bound together by Daniel's story: communication and ethics. Both are necessary for an end-of-life discussion. Without full and ethical communication, a heavy burden is placed on the patient, the caregiver, or the family to make decisions that are not necessarily in the best interest of the patient.

In the chapters to come, we will at times separate these two strands in order to dissect them, to tease into the open their individual concerns, but that is a merely a paper ploy. Good communication is always ethical, and ethics depends on communication.

Reference

Institute of Medicine. 2014. *Dying in America: Improving Quality and Honoring Individual Preference Near the End of Life*. Washington, DC: National Academies Press.

Defining the Patient Population

2

Abstract

The ethical process is not a solution to communication gaps but a variation of mediation tactics, listening, and a value-based perspective. Such communication might be a good protocol to replicate in the clinician–patient/surrogate relationship. This original research includes data from ethics consults taken from 560 case studies in hospitals from 2008 to 2012; eventual results were analyzed as to whether patients responded positively or negatively to the recommendations. The type of person (age, race, gender) requesting an ethical consult was reflective of the overall hospital population. However, when end-of-life cares were concerned, researchers found that an element of miscommunication was rooted within the healthcare team; that miscommunication lay at the heart of every patient request for the consult, no matter the level of socioeconomic and demographic status for the patient. Correlation was found between delay in requesting an ethics consult and negative outcomes of such consults (i.e., patients not following recommended actions). Correlation was not found for age, race, gender, or any other factor. In addition, 90 percent of ethics consults were requested by surrogates or clinicians rather than patients, an indication of a highly vulnerable group.

Most disagreements that occur in ICUs are a result of communication issues between clinicians and patients/patient families. Clear communication has been shown to reduce time needed for decision making and increase family satisfaction. Most disagreements between physicians and decision makers can be resolved with "proactive communication." Proactive communication = physicians listening to families, asking about patient wishes, explaining a prognosis without confusing medical jargon, and discussing treatment plans in the context of a patient's wishes and quality-of-life concerns. It must be reiterated that these are a small sample of case studies over a fairly short period of time. Results may not necessarily reflect those that might be found in a larger study. The author encourages other researchers to conduct similar studies to provide a more complete picture of communication studies.

© The Author(s) 2017
K. Benton, *The Skill of End-of-Life Communication for Clinicians*,
SpringerBriefs in Ethics, DOI 10.1007/978-3-319-60444-2_2

Keywords
Ethics • Physician communication • Ethical dilemmas • Palliative communication

My brother Daniel, who experienced lifelong suffering from the elephant man disease, represents a forgotten population in the clinical realm. He is one of the ones who will die. His disease is terminal, but ultimately at some stage in life and in disease, he is everyone. Without needed citation or reference, I can with certainty write that 100% of the human race will have 100% mortality at some point in time. Clinicians must not abandon, ignore, or be underskilled in the area of dying.

Many studies deal with communication issues, particularly in hospitals (refer to the list at the end of this chapter for particular sources). In this chapter, however, we concentrate on 560 cases, most of which concerned end-of-life issues. And because end-of-life issues naturally affect an older population more directly rather than those of Daniel's age, we emphasize those sorts of concerns. The need for this communication, however, spans all ages.

More so now than ever before, many countries are composed of an aging population and a more chronically critically ill subset of citizens; sustained by treatments and advances, people are living longer. Understanding this population of people in need of medical care, wherever they are based, is also crucial. The cultural norms and understood truths of medicine vary from community to community and in hospital institutions throughout the inhabited world. Familiarizing oneself with one's surrounding population can shape the chosen "best" skill for communication. Resonating with the lessons learned in patients' stories is vital to the value of empathetic treatment.

In particular, the skill of communicating about end of life is not regularly taught, although it is just as unique and intricate as any other procedure, code, or medication provided. It may have gone largely unnoticed until now because in planning a medical curriculum no one aligned the action of *talking* with a procedure or skill. It is only recently that the skill has been given its own billable code in the United States, designated *advance care planning*. The sad truth is that clinicians in general need training in communicating as much as they do in operating. One seems innate and the other is learned, but which is which depends on the profession and the person. There are ways to approach patients and ways to avoid approach. Each must be understood and become second nature. To have nothing more than a stale set of questions to rattle off is not to understand the skill that people, our patients, deserve..

We are a culture of elderly and their caregivers. In the United States and in other parts of the world, our baby boomers are aging and their parents are still living, just older. Thus, many of us are part a culture of the chronically ill. We are a culture of artificially supported patients, dependent on a variety of technology that will support life for many years, but machines have a cost, both in terms of actual dollars and in how they complicate care, enhancing chronic illness. In most countries, payment structure does not promote or incentivize comfort care at the end of life. Payers allow continued and ongoing end-of-life care with no limitations other than pro-nouncement of death. Therefore, no matter how inappropriate or appropriate, bur-densome or beneficial, the treatment mechanism is, it will be paid for and allowable if ordered by the physician, consented by the patient or his/her surrogate, and administered by the team. This leads to a number of disconnects when the benefits of a particular course, therapy, or piece of equipment provides no basis for increased revenue production. Our focus, therefore, must shift to learning about patient dig-nity and best interest because it is the constant in all cases, as well as the ethical thing to do.

Depending on the insurance agency, the hospital, the state in which the patient lives, and a whole host of other variables, most of the in-hospital treatment is paid for (at least in part) as well as much of the at-home treatment (the medicines, the equipment, the medical professionals, and so forth), but the caregivers are not, at least not to fill a 24/7 shift week. The caregiver tends to be an underpaid, seldom paid, or not a paid field. Caregivers are emotionally bound, exhausted, and over-whelmed. So in addition to needing to communicate with in-hospital professionals, another layer of communication is needed with regard to at-home treatment; these caregiving individuals are those with whom you need to communicate when the patient is not cognizant or is overwhelmed by emotion. If poor communication is at the root of many ethical dilemmas as noted in the many case studies here, let us not forget this secondary audience as another factor to be recognized, addressed, and overcome.

2.1 Value-Based Concerns

Palliative communication is a much richer process than many other clinical conver-sations. Whether presented by the palliative team or the care team, the palliative ideology embraces the whole patient being, beyond the one failing or improving organ. Ethically one must honor the wishes of the patient as he or she expressed them. After a long illness, some patients no longer want to hold onto this life, expecting a better life after this one. Others are not yet ready to give up their earthly presence. Religion (no matter which one, or a lack of religion) plays a strong part in this decision. Each case, each set of communications and ethical decisions, presents a different set of variables—but in every case, ultimately what matters is what is best for the patient as he or she sees it. Input from the doctor, the family, friends, religious figures, and everyone else the patient considers important are vital to this decision, but it is still the patient (or the surrogate) who must decide.

Barring accidents or sudden events, a person's end of life is generally a phase that may last for hours, days, months, and in some cases, years. In many ways, the longer the phase, the easier it is for both patients and their families to come to terms with what is happening. Yet death, however long it may have been anticipated, is always a shock to those still living. It is human nature to wonder what might have been, if only we had done this other thing differently. When a patient is open and honest about what he or she wants, when the patient knows the full facts of his own case, when the teams work together by communicating the possibilities, the potentials, and the perils, those still living can know they have done all they could for their loved one. The key in all cases is communication, even about death—especially about death.

2.2   Ethics and Communication

There is a clear alignment between the relationship of the ethical dilemma in clinical practice and communication conflict. This became clear when my team of researchers looked at a variety of causations for consult with someone in the Ethics Department. What they found was that some form of miscommunication led to the majority of requests for consult, tying up hours of professional time and resources.

Ethics at its very root is simple mediation, negotiation, objective review, and advisory support. Put simply, one might define the process as a form of artful communication. In the cases statistically analyzed below, the ethics consult drastically changed the care plan decision in over 80% of cases, almost always with a better result for the patient. The action to catalyze the change was the simple task of communication. The hindered communication either between a patient and family or care team and patient/family was reflected in the reason for consult, whether it regarded an advance directive (AD) issue or the need for surrogate consent. Among other things, the study found that the type of person (age, race, gender) requesting the ethical consult was reflective of the overall hospitals' populations. However, when end-of-life cares were concerned, there was an element of miscommunication rooted within the healthcare team no matter the level of socioeconomic and demographic status for the patient.

Although the field of medical ethics established concepts such as patient autonomy, informed consent, and confidentiality, clinical ethics applies those paradigms to real-time practice. Overall, clinical ethics exists to bridge the gap between laypersons and clinicians when reconciliation seems unlikely without a neutral intervention. Many hospital systems and outpatient facilities do not use the support of an ethics service partially because it is not mandated, and partially because it might be viewed as a noted intrusion in the doctor–patient relationship. However, clear communication—with or without an Ethics Department—is necessary for an ethical solution to end-of-life issues. The ethics role brings a fresh set of eyes and a patient advocate perspective. Despite the population present in a hospital census, if a clinician is able to empathize with the patient and yet provide ethical solutions to the problem of extending treatment in a futile setting, for example, the end-of-life dilemma of when to stop treatment ceases to exist. In addition, clear communication has been shown to reduce the time needed for decision making and to increase family satisfaction (Bosslet et al. 2015). This ethical analysis, though successful with the use of an engaged ethicist, could change the role of the Ethics Committee if it was viewed as an outward template for other staff to emulate all of the time, and to use as their own. Theoretically, this might lead to fewer Ethics consults and more straight-to-the-heart discussions by caring professionals and their patients.

Understanding the results of a population analysis can give a clinician a sense of best practice skill in choosing a communication strategy, whether that population is older or younger, religious or not, well-to-do or not so well off, predominately one gender or another or both, and so forth. The value of knowing a population remains self-evident.

Knowing what families go through, as we experienced with Daniel, I wanted to make sure such cases of miscommunication were eliminated as much as possible. This was my hope as a sister. As an ethicist with more than a decade of experience in professional, clinical cases across the United States, I seek to change the trend and improve the communication skills of all clinicians with regard to end-of-life discussions. The ethical process is not a solution, but a variation of mediation tactics, listening, and value-based view of communication which might lead to a good protocol to replicate in the clinician–patient relationship.

My research utilized ethics cases from 2008 to 2012 for statistics. The 2013–2016 years were then qualitatively analyzed for anecdotal examples on lessons learned and best practices (to be discussed in the last two chapters of this book). Data for ethics consultations were maintained on a rolling basis in narrative summaries (Microsoft Word) as well as spreadsheets (Microsoft Excel). Quantitative and demographics data (admission date, age, gender, race, existence of advance directive (more commonly known as a healthcare proxy and living will), vent status, code status, date of consult request, and reason for consult) were obtained from electronic medical records or from the patient's chart. Qualitative data such as reason for consult, surrogate, committee recommendation, and outcome were pulled from written narratives or electronic ethics notes and categorized by coded theme. Consults were uniquely identified by case numbers unrelated to the patient identifier in the hospital system to maintain privacy.

Each year was analyzed to determine the number of cases, which were then fixed to determine percentages and means for various categories. The five-year case total was 560. Results were reported as mean, median, range, percentage, or real numbers, as appropriate. Summative results for the five-year period were also calculated for most categories. Reasons for consults were coded according to a thematic list. Instead of focusing on specific outcomes, the results of the cases were determined as positive or negative based upon decisions made for care plan. Positive outcomes were defined as decisions made or actions taken that were deemed by the committee as being beneficial for the patient. Negative outcomes were defined as measures taken that were not deemed in the patient's best interest, such as refusing comfort at discharge with concomitant readmitting soon after. Since these actions are patient- and family-centered endeavors, the outcomes measure was based on the choices made and action taken by the both parties.

2.3   Results

Basic demographic data (sex, race, age) appear in Table 2.1. Data for each year are presented individually with a five-year summative total at the bottom. The small difference between the mean and median for each year indicated that the age distribution for consults was approximately normal with a tendency to skew toward the older ages, which was not unexpected. Overall, we did not see any indication of an age bias and concluded that patients of any age may request an ethics consult. Gender bias was also not evident given that the means closely aligned with local census data. The race mix changed over the interval, going from strongly favoring African American over Caucasian in 2008 to more closely representing the overall environment census by 2012.

Table 2.2 represents data characteristics that are related to hospital status, including mechanical ventilation status, medically assisted nutrition and hydration (MANH), and if a patient was located in a critical or intensive care unit. These data were computed as percentages of the total number of cases for each year with five-year summative means.

Table 2.1 Gender, race, and age demographic summary for ethics consultations 2008–2012 (numbers rounded)

Year	%Male	%Female	%Cau	%AA	%O	Mean age	Median age	Range
2008	52	48	39	59	2	64.0	63.0	21–96
2009	48	52	40	54	6	62.9	64.0	22–100
2010	48	52	44	52	3	65.6	66.5	23–100
2011	44	56	46	50	4	68.5	70.0	23–99
2012	52	48	51	43	6	67.0	69.0	22–98
Five-year mean	48.8	51.2	44	51.6	4.2	65.6	66.5	–

Note: *Cau* Caucasianm, *AA* African American, *O* other races. Race codes from patient medical records were used

Table 2.2 Hospital-related demographic data for consults

Year	# of cases	On Vent (%)	MANH (%)	Location (%)
2008	88	26[a]	24[a]	67
2009	102	39	33	63
2010	126	40	13	56
2011	102	25	24	54
2012	142	46	23	75
Mean	–	35.2	23.4	63

Note: On Vent–Patient requiring ventilator assistance for breathing at time of consult request
MANH–Patient on medically assisted nutrition and hydration at time of consult request
Location–Patient in a critical or intensive care unit at time of consult request
[a]Data incomplete for the year—28 case data sets were missing

Table 2.3 Summary of reason for consultation categorically by year with five-year summation/ percentage of total cases

Reason	2008	2009	2010	2011	2012	Total	% of total cases ($N = 555$)
Family issue	10	18	24	16	61	129	23
Decision making/Pt competency	17	17	43	22	20	119	21
Code status	9	15	33	14	17	88	16
Futile care	23	17	5	6	3	54	10
Advance directive	3	8	2	9	29	51	9
Surrogate issue	17	16	1	8	0	42	8
Plan of care	2	4	14	11	3	34	6
Doctor issue	3	2	0	5	4	14	3
POA issue	1	1	2	7	1	12	2
End-of-life issue	2	1	0	2	2	7	1
Gestation/fetal issue	1	3	0	1	0	5	1

Note: $N = 555$ indicates exclusion of five cases due to lack of data pertaining to this category

Table 2.4 Advance directive (AD) and DNR (Do not resuscitate) status

Year	Total consults	AD (%)	DNR (%)	AD/DNR (%)
2009	102	10 (10)	18 (18)	3 (3)
2010	126	23 (18)	7 (6)	4 (3)
2011	102	33 (32)	21 (21)	11 (11)
2012	142	27 (19)	65 (46)	14 (10)
Total	472	93 (20)	111 (24)	32 (7)

Note: AD/DNR indicates concurrent DNR with the existence of AD. Data for 2008 were not available. Data indicate the real numbers for existence of DNR or AD at the time the consult was requested

Reasons for an ethics consult request were either pulled from the electronic record of the request or from the ethics narrative record. In the event the reason came from the narrative, the overall theme was summarized in the spreadsheet and subsequently coded. Table 2.3 summarizes the reasons for consults by year by case count, with summative case counts and percentages of total cases. It should be noted that only the primary reason for request is listed; the actual situation often had multiple issues to address once the ethicist evaluated the issues.

Table 2.4 portrays the presence of an advance directive (AD) or a Do Not Resuscitate (DNR) order in place at time of consult request. Status was determined either from the admission information or from the patient's chart. Concurrent presence of AD and DNR was also determined.

Table 2.5 represents the interval between patient admission and request for an ethics consult. Cases were categorized as days postadmission until consult request and intervals set at 0–5 days, 6–10 days, 11–15 days, 16–20 days, and ≥21 days. Overall, 45% of cases were called within 5 days of hospital admission and 76%

Table 2.5 Number of cases per day-time interval postadmission to consult request by year with five-year summation as a percentage of total consults ($N = 555$)

Year	0–5	6–10	11–15	16–20	21+
2008	41	16	11	5	13
2009	38	24	12	6	21
2010	45	34	12	11	23
2011	52	22	9	6	12
2012	76	21	13	7	25
Total	252	117	57	35	94
% of total consults	45	21	10	6	17

Note: 2008 lacked data for two cases; 2009 and 2010 each had one case that was not carried to full consult; 2011 lacked data for one case

Table 2.6 Negative or no resolution outcomes by days postadmission

Year	0–5	6–10	11–15	16–20	21+
2009	5	5	2	2	7
2010	5	8	0	1	5
2011	14	5	0	1	4
2012	9	2	1	2	9
Total	33	20	3	6	25
Total cases per interval	211	101	46	30	81
% of cases	16	20	7	20	31

Note: 2008 excluded due to lack of data regarding outcomes

within 15 days of admission. Consults were requested 17% of the time at 3 weeks or greater.

Table 2.6 displays the number of negative outcomes by year per interval with a summative percentage. Reason for the consult is commonly due to family issues, decision making, code status, or futile care issues. Although this is an accurate summarization of the "typical case," it lacks insight into the population and why an ethics consult was needed. To go deeper, we will take an epidemiological look at the demographic categories individually.

2.4   Conclusions

No real trend regarding ventilation status or patient location is evident. The presence of a concurrent DNR with an AD is low, meaning that a patient's living will (AD) translated to a wish to allow a natural death did not prompt a DNR order unless a lengthy discussion took place. This is worrisome since many lay patients and professionals falsely assume once they complete the AD, their DNR will be put into place at end of life. We did note the tendency of a DNR in place at the time of consult increased over the interval. This upward trend is attributed to increased staff education and the willingness of providers to discuss DNR status with families prior to requesting a consult.

The primary reason for the need of an ethics consultation can be difficult to determine. Communication with multiple clinicians and shift changes can lead to confusion and misunderstandings, requiring solid communication to sort through and determine underlying issues. Complicated family structures may require identification of the legal and/or de facto decision maker. Ethics consultants must be able to discern the dynamics within a family group to determine what the real issues are and how to approach families with solutions that benefit the patient. Discussion about DNR or hospice may upset the patient who does not truly comprehend what those entail, making them highly resistant to such decisions. Fundamentally, the need for an ethics consultation often results from a lack of communication and solid information.

Approximately 90% of decisions made within this population were made by surrogates rather than the patient, indicating an extremely vulnerable patient population. The ethics consultant would attempt to communicate with the patient directly in each case; however, patients often were physically and mentally unable to communicate. Since the patient is in such a vulnerable state, determination of the true legal decision maker is vital.

A trend of particular concern is the catalyst to negative outcomes. Data indicate that while 17% of total consults are called at 3 weeks after admission or longer, this time interval is responsible for 31% of negative outcome cases. The time interval of 16–20 days produces 6% of consults, but 20% of negative outcomes. The zero-to-five-day interval has the highest number of consults with the second-lowest negative outcome percentage. The trend indicates that the greater the number of days since admission to consult request, the higher likelihood of a negative outcome. Earlier intervention appears to produce better outcomes for the patient. This area requires further study.

In summary, involving Ethics produced a positive outcome for patients 82% of the time. This means 82% of the time patients transitioned to comfort and less aggressive measures. Postive outcomes define an amiable and mutual decision made by patient/surrogate and care team. An Ethics consult at its root is a process of bringing loved ones together, of reviewing all clinical and nonclinical information, and hours of conversation to ensure understanding of prognosis. The most successful outcomes occurred early in an admission. The Ethics involvement begs the question: if most clinicians were claiming good communication, how is the Ethics consult producing different outcomes? Before Ethics comes in, the physicians claimed that patients and/or surrogates demanded futile treatment and were uninformed. Once Ethics became involved and open communication established, beneficial results occurred.

Learning how to communicate with patients, families, and caregivers is crucial because it can act as the solution to the myriad problems in health care at end of life for all concerned. Somehow, someway, by implementing all of the advancements that have beautifully led to longevity and healing, we have dehumanized and medicalized end of life. So let us next evaluate the ways in which we might rehumanize this area through the art of communication.

References

Back, Anthony, Robert Arnold, Kelly Edward, and James Tulsky. VitalTalk. Available at www. vitaltalk.org.

Bosslet, Gabriel T., Thaddeus M. Pope, Grondon D. Rubenfeld, Bernard Lo, Robert D. Truog, Cynda H. Rushton, J. Randall Curtis, Dee W. Ford, Molly Osborne, Cheryl Misak, David H. Au, Elie Azoulay, Baruch Brody, Brenda G. Fahy, Jesse B. Hall, Jozef Kesecioglu, Alexander A. Kon, Kathleen O. Lindell, and Douglas B. White. 2015. An Official ATS/AACN/ ACCP/ESICM/SCCM Policy Statement: Responding to Requests for Potentially Inappropriate Treatments in Intensive Care Units. *American Journal of Respiratory and Critical Care Management* 191 (11): 1318–1330.

Lakin, Joshua R., Susan D. Block, J. Andrew Bilings, Luca A. Koritsanszky, Rebecca Cunningham, Lisa Wichmann, Doreen Harvey, Jan Lamey, and Rachelle E. Bernacki. 2016. Improving Communication About Serious Illness in Primary Care. *JAMA Internal Medicine* 176 (9, July 11): 1380–1387.

Lowery, Susan E., Sally A. Norton, Jill R. Quinn, and Timothy E. Quill. 2013. Living with Advanced Heart Failure or COPD: Experiences and Goals of Individuals Nearing the End of Life. *Research in Nursing and Health* 36 (4): 349–358.

Ethics and the Medicalization of Death

3

Abstract
End-of-life care is not based on research-tested, evidence-based protocols because that research does not yet conclusively guide physicians to withhold care with absolutely certainty of the burden. According to Scott D. Halpern in the *New England Journal of Medicine*, the announcement by the Centers for Medicare and Medicaid Services regarding reimbursement rates for EOL care planning is a step in the right direction. However, the published research setting absolute standards for artificial support might change the culture dramatically. To determine which interventions actually improve outcomes, "more frequent and rapid conduct of large randomized trials and quasi-experimental studies" are necessary (p. 2001). A consensus needs to be reached about which outcome measures should be used in the studies to measure EOL intervention success. The urge to "do something" is leading healthcare organizations to implement policies and approaches that have not been tested. Slowing down and pursuing education and research before implementation can help achieve more successful outcomes in EOL care. EOL communication needs to allow patients and families to make fully informed treatment decisions that will lead to more successful long-term efforts and more telling results about quality outcomes.

Keywords
Ethical principles • Ethics of end-of-life communication • Advance care planning

The medicalization of death is, in essence, the overtreating or treating to death of a terminally ill patient. There is a vicious cycle that begins slowly at the onset of a chronic illness and extends to the end of life. This cycle is at the heart of what makes modern medicine both brilliant and burdensome. No matter where a hospital is, with an aging population, more people will live longer, the majority will live to an older age, and many will have chronic illnesses for a longer period of time. A patient

K. Benton, *The Skill of End-of-Life Communication for Clinicians*,
SpringerBriefs in Ethics, DOI 10.1007/978-3-319-60444-2_3

might be diagnosed with chronic obstructive pulmonary disease (COPD) and therefore might need to seek specialty care and eventually long-term O_2 or even BiPap support, for example. These patients' needs for specialists and aggressive measures to treat their illnesses may lead to repeated infections, issues with nutrition, and, eventually, chronic hospitalizations. Frequent emergency room visits with repeat admissions or long-term acute care (LTAC) facilities are where your patients may go when they are too sick to discharge home or to a nursing home, are possibly still on support, and are simply awaiting death, and yet no one—not family, not caregivers, not clinicians, possibly not even the patient himself—wants to give up hope or to admit that further treatment is likely to be futile. But is this the most ethical course of action? Four moral principles of bioethics lay the groundwork for ethical decision-making: autonomy, beneficence, non-maleficence, and justice (Head 2011).

In an article written about LTACs, they are referred to as "the perfect storm" because antibiotic-resistant infections run high due to a population of compromised patients (Gould et al. 2006, 924). The patient returns to the acute hospital with a multitude of infections or vicious pain, only to be discharged for days or weeks. It can be argued that these are not good places for anyone to experience a peaceful end phase, and yet many do. Insurance reimbursements decrease to hospitals when this cycle of readmission occurs, but instead the obvious is left unstated: this cycle is the circle to death that cannot be treated well or healed, not even by the most well-trained physicians and staff.

This chapter has two main goals: (1) to understand how we medicalize death, in purely ethical terms, and (2) to understand how we undo medicalization of death through good ethics—that is, good communication combined with value-laden discussions, with a focus on dignity over documentation—very simply, informed choice and patient-centered best interest.

3.1 Principles of Healthcare

So why do we potentially overdo treatment? Let's first look at how our basic ethical principles play into this venture.

3.1.1 Autonomy and Paternalism…a Balance

If we are in a mostly independent and self-directed culture, why do we not subscribe to the true informed right of the autonomous individual? We allow patients and their families to demand everything and anything at the end of life, but our present hospital and caregiving mind-set does not ensure that those responsible for making medical care decisions are informed. We have done our patients a disservice by allowing them to direct their own care, carefully excluding ourselves from the liability of decision making. Self-law (autonomy) means self-directed and educated, not just choice. Since autonomy is by definition self-oriented, we have no guidelines or requirements for how to deal with autonomy in others.

And what is our clinical autonomous responsibility? That is a difficult concept when life has reached its end and a menu list of technology is available to prolong being alive, maybe through burdensome care, maybe through aid. The right to choose is an absolute and should never be replaced, but I make the argument that the demand to offer every choice does not contribute to autonomy. For example, an oncologist once argued that all his patients were well informed after he offered them a compassionate clinical trial, despite being at an end stage of their cancer. I asked the oncologist, "How many said no?" "None," he responded. But is this a fair evaluation when none of these patients were in a position to clinically evaluate their own status? Most of them undoubtedly knew that the risky trial would not cure their disease, yet still they hoped, despite the pain and suffering the trial would bring, and would help them prolong their lives.

Likewise, much of the time in a hospital setting, the "autonomy" comes directly from the patient's surrogate. The concept of a healthcare proxy hinges on the decision maker's ability to suspend all biases and make choices based *solely* on the patient's wishes. Healthcare proxies are expected to process complicated medical information under a stressful situation and still make choices that benefit the patient and respect his or her wishes. Clearly, the expectations for decision makers are high, and several common mistakes have been identified that prevent proxies from making the best choice.

Proxies often override a patient's DNR wishes and choose to have the patients sustained on life support to "save" the patient. In that case, the patient is forced to live with a disease and in a condition that goes against his or her wishes.

Another mistake decision makers can fall into is making choices that reflect a sense of duty. Proxies will choose to sustain a patient on life support even when it is clear that the patient is suffering. When asked why the proxy is making decisions to keep a loved one on life support, the decision maker often expresses a sense of duty (Periyakoil 2015).

In the food industry, we allow autonomous choice when a patron chooses a meal at dinnertime, but the facility is not forced to add raw chicken to the menu simply because it is back in the kitchen. Everything should not be included on any autonomous menu. Paternalism—a medical and ethical term referring to the physician's duty to guide and invoke choice for the patient—is a possible balance to informed autonomy. The practice of paternalism has dwindled as years have passed, more prevalent prior to and during the 1950s. Today, during patient–clinician communication, physicians often hide behind the guise of professionalism to remain detached from patients. All the bedside, housekeeping, and medical guidelines and rules about how to interact with patients may actually be hurting a physician's ability to communicate and care for patients. Patients and families want to connect with healthcare professionals, and often that requires doctors to find common ground in what might be considered unprofessional ways, such as talking about their late nights out, finding commonalities, sending a card. Some risk is involved, but speaking with families in a personal way can help physicians to better care for patients and understand what they truly need as opposed to a medical imperative to prolong life for no other reason than to do just that (Javier 2016).

When patients are seen by a battery of specialists (hospitalist, oncologist, pulmonologist, etc.), who is responsible for telling the patient that he is dying? With Medicare reimbursing for EOL conversations, someone has to take ownership and be the one to tell a patient that he has crossed the threshold into EOL. Making advance care planning and EOL conversation billable should encourage clinicians to hold each other accountable and force someone to take the time to fully explain EOL options and patient wishes (Nuila 2016). Many times I have listened as families ask me why the physician will not bring up the end or "let go of the treatment." That is a sad twist of reversed roles that needs an immediate fix. Thus, you give value to autonomy and paternalism; we find balance by offering viable options and steering clear of those that will *only* burden, even if longevity is attached to that burden.

3.1.2 Justice

Justice is the second basic ethical principle. Justice has become a dirty word in American healthcare when aligned with treating the last stage of life. In a just world, everyone deserves everything without consideration of diagnosis or prognosis. But let us consider the possible injustice this is doing to those among us when we treat in a way that has no chance of curing the patient. Is it just to burden the body so that we know everything has been done for everyone, perhaps only to relieve an anticipatory guilt? In its 2015 report, the Medicare Payment Advisory Commission (Med PAC) notes that for roughly two decades, healthcare costs in the last year of life have consumed more than 25% of all Medicare spending. That statistic has not changed (Levins 2016). That is an injustice when so many are without basic preventative care and die of unnecessary cause. The true justice left in the end of life is the promise of mortality: no one has yet avoided the expectation of death. Life is the terminal illness. There was a point in history where death was not so marginalized and was viewed as a natural progression to repopulate and continue the circle of life. Somewhere along the way, we have chosen to fear death and to avoid it, as if by denying its existence, death itself might not happen.

In the work I do with the indigent and uninsured population on the outpatient level, we are able to look at true justice and best interest for the patient because these patients do not have the liberty of seeing a multitude of specialists, due to the missing payer source. Seeing many physicians can be a double-edged sword and is sometimes the recipe for lost communication and incongruent treatment paths. To avoid this, one specialist might agree to the charity care and guide the rest in considering the treatment path using telemedicine in the clinic. Likewise, a palliative discussion is immediately offered and a boundary-setting discussion is held, which allows the patient an early chance to set his or her preferences. Interestingly, in pursuing the care in this respect, these indigent patients receive better and more just care than their insured counterparts. They are not lost in the complexities of specialties; they are clear on primary, secondary, and eventual end-stage goals. These goals might be reflective of wanting "everything" done, and that is ok and just. What is important is the ability to communicate every scenario, logistical barrier, and reality of care that might ensue, so that they know.

3.1.3 Best Interest or Beneficence

In Ira Byock's book *The Best Care Possible*, he writes of the importance of practicing loving care. This is the care that humanizes the treatments and technology, and truly offers healing, whether there is option to cure or not. There is always opportunity for healing and for hope. Medical technology designed to prolong life is often used in a way that disregards the quality, dignity, and humanity of that life. Physicians and patients have become accustomed to having numerous options, and EOL care is no exception. Patients and families discuss treatment plans but often fail to discuss the quality-of-life limitations associated with much of the life-prolonging medical technology that has become commonplace. As a result, many patients are being put through long and uncomfortable deaths in the hospital, dependent on machines and unable, unwilling, or unaware of other options (Profeta 2016).

How can one know another person's best interest? This can only be defined by the patient and the patient alone. Providers cannot vouch for knowing a patient's best interest. This is why a conversation about what the patient needs and wants must begin early and be ongoing (Levins 2016). There is little research to guide doctors and nurses in deciding when and how to pursue end-of-life measures for their patients. No policy currently exists to help clinicians choose appropriate end-of-life or palliative measures.

The problem is compounded by the fact that most healthcare professionals avoid broaching the subject of death with patients and their families. Nor are health service researchers interested in pursuing investigative efforts intended to shed light on end-of-life interventions and outcomes. Most researchers cite a lack of return on investment as the reason that end of life is such a neglected research topic (Levins 2016). Palliative care is certainly a solution aligned with a patient's clear best interest, but as in Daniel's case, it can still be wrought with negative impact when a stranger begins a discussion that should have started with a familiar provider. Palliative MDs are strangers much of the time. Because these strangers lead the palliative care discussions, increased stress when talking about goals of care for all families of patients with chronic illness may occur (Carson et al. 2015). If the familiar doctor can perform the same communication function or at least segue into the relationship with the palliative team, the benefits seen through patient satisfaction may be measurable.

"Best interest" is defined from the point of view of the patient and included in that best interest must be a basis of *informed* interest. Most people experience debilitating pain, confusion, depression, and shortness of breath in the last 12 months of life. But when there has been no or little advanced care planning discussion, medical treatment in the last year of life often violates patient wishes. Because there is such a huge deficit of palliative care professionals responsible for EOL communication, all physicians should be trained how to use established best practices for communicating prognosis, planning goals of care, and managing symptoms. The important focus may not be on a good death for patients but instead on a good life up until the very end. Physicians need to be able to provide a good quality of life for patients with life-limiting illnesses (Gawande 2016).

3.1.4 Do No Harm

Providers and allied professionals alike take a vow to do no harm, but do not always consider the care as the actual harm. (Note: Although the original Hippocratic Oath never carried the phrase "Do no harm," the phrase is a brief précis of the oath.[1]) If the goal of treatment is to heal, families of patients given a poor prognosis have trouble making clear, informed decision about their loved one's care for many reasons; physician communication issues are only part of the problem. First, patient families suffer from an "optimism bias" (Span 2016, D5) that makes them more likely to interpret a poor prognosis with optimistic predictions. Family member/decision-maker beliefs also skew decision-making ability and prognosis understanding. Mandating physician communication surrounding dying may bring the benefit of allowing patients to make fully informed and deliberate decisions about end-of-life care. After passing the legalization on physician-assisted suicide in Oregon, opponents of the legislation argued that palliative care efforts to keep patients comfortable would decrease. As is evident in Oregon, the quality of palliative care has actually improved. Other physicians (besides only palliative professionals) have been motivated to better educate themselves on palliative care measures. In a state where doctors are faced with a high likelihood of end-of-care conversations, Oregon's healthcare professionals are spending more time with patients and having thoughtful conversations about EOL care options and patients' wishes (Lindsay 2009). Why can't this conversation improve without the need for a patient to request a physician-assisted death? Why not improve the conversation surrounding suffering first?

Decision makers, who often consider positive thinking or "beating the odds" by deeming their loved one a fighter, will help push the prognosis in a positive direction (Span 2016, D5). Still, a frank and honest discussion with a physician about prognosis serves to help the decision maker. Even though there is skepticism and doubt on the part of the decision maker that a physician can accurately predict patient outcomes, families and patients want to make decisions based on a transparent conversation with the doctor. Families can more easily plan for end of life once a doctor can assess and verbalize that a patient's condition is terminal (Span 2016).

---

[1] A portion of the modern version of the Hippocratic Oath includes the following (emphasis added): I swear to fulfill, to the best of my ability and judgment, this covenant: … I will respect … and gladly share such knowledge as is mine with those who are to follow…. I will apply, for the benefit of the sick, all measures which are required, avoiding those *twin traps of overtreatment and therapeutic nihilism*…. I will not be ashamed to say "I know not," nor will I fail to call in my colleagues when the skills of another are needed for a patient's recovery…. Above all, I must not play at God.

3.2 The Details in the Tools and Documents

3.2.1 The Truth About the Advance Directive

In order to seek a solution to the ethical conundrums that this chapter has explored both philosophically and applicably, it is important to be knowledgeable about the strategies and tools that do exist to aid your practice.

First is the theoretically well-known advance directive, commonly aligned with the power of attorney for healthcare and the living will combined, sometimes called "five wishes." It is, in short, an essence of what a patient has written to express who is appointed as healthcare decision maker when she cannot speak for herself, and to say what she agrees with or limits regarding treatment at end of life, *and end of life only.* On a personal basis, mine says to allow natural death, but I still want to be coded because I am not yet designated end of life. What this document *is not* is a Do Not Resuscitate (DNR) order. Writing "no machines" does not mean I refuse machines at all times and under all circumstances; it simply means I plan to refuse those mechanisms once I am at the end of life, period.

3.2.2 POLSTs and Other Documents

A POLST or MOLST document (Physician Order for Life Sustaining Treatment) is a standing physician order that makes a patient, by consent of the patient or proxy, a DNR at that present point in time. These documents are legal in many states and provide conversational guidelines, spelled out in a transitional order form, that acceptably may move from facility to facility to home and nursing home with standing orders for end-of-life wishes. It would not be appropriate for a well person to have a POLST completed. Unfortunately, DNR and all the documents that support the idea are frequently misunderstood by the lay and the healthcare world alike.

When compared to medical students and other healthcare workers, doctors and nurses have different preferences about cardiopulmonary resuscitation when faced with a terminal illness. In the event they are diagnosed with a terminal illness, doctors and nurses are more likely to change their code status to DNR. This might indicate that doctors and nurses have a more realistic view of terminal illness and the outcome for themselves. However, the general public and less-educated healthcare workers share a more optimistic view of terminal illness and the impact of cardiopulmonary resuscitation. When speaking with a patient's family, many doctors adopt this more optimistic view in order to avoid conflict between patient perspective and physician recommendation on DNR status in the event of terminal illness (Chavez et al. 2015).

Clarity on resuscitation orders and end-of-life care options are critical for healthcare teams to understand, and very unfortunately the gap in knowledge is prevalent. We cannot expect patients to understand de-escalation and boundaries if nurses, therapists, and providers are misinformed. Palliative care differs from hospice in that it is appropriate "at any age, for any diagnosis, at any stage in a serious illness," and it can be provided in conjunction with disease treatments (Meier 2016, slide 23). Patients do not have to discontinue other treatments to receive palliative care. Palliative treatment is designed to improve quality of life for the patient, which often involves cost reduction through fewer hospital admissions. Patients can often stay home and out of 24/7 care facilities with the help of palliative care. Research supports the assessment that palliative care reduces the likelihood of medical error by increasing quality and length of the patient's life, increasing patient satisfaction, reducing major depression, reducing "aggressive" treatments, and reducing the likelihood of unwanted or overly burdensome treatments. Palliative care can help connect families to existing services and support caregivers (Meier 2016).

When palliative services are not available and staff is not adequately trained or comfortable, there are tools to assist communication. This is that point in time where we reach even further in lieu of avoiding the elephant in the room, such as those introduced and included in Angelo Volandes's *The Conversation: A Revolutionary Plan for End-of-Life Care*. His team created a video library that clearly interprets, explains, and provides good information on the concept of extraordinary measures and limiting these measures. These videos can be found at https://www.acpdecisions.org/patients/. A multiyear study in Hawaii is proving the validity of the use of the videos. Evidence-based articles to support this movement can be found at https://www.acpdecisions.org/evidence/. Early implications of a recent study by Volandes et al. indicate that "implementing ACP [advance care planning] video decision aids was associated with improved ACP documentation, greater use of hospice, and decreased costs" (Volandes et al. 2016, 1035). Studies such as this one should stimulate further research into end-of-life care planning, where there are benefits for all: patients (better quality of life), hospitals (decreased costs), and physicians (clearer communication).

Another invaluable resource is the website of the Conversation Project, where providers and care teams can find information on how to jump-start the necessary dialogues with patients and/or families (http://theconversationproject.org/).

A professionally qualified palliative care team, found throughout most of the United States and at many locations internationally, is another overlooked resource and can help the dialogue between provider and patient—but even the best of these professionals cannot replace the need for a strong rapport between a familiar provider and a patient. Their research offers support for improved communication, but the anecdotal narratives will move and prove the true necessity if we are to adhere to value-based ethics and absolute compassion for our ill patients.

References

Byock, Ira. 2012. *The Best Care Possible*. New York: Penguin Group.
Carson, Shannon S., Christopher E. Cox, Sylvan Wallenstein, Laura C. Hanson, Marion Danis, James A. Tulsky, Emily Chai, and Judith E. Nelson. 2015. Effect of Palliative Care–Led Meetings for Families of Patients with Chronic Critical Illness: A Randomized Clinical Trial. *Journal of the American Medicine Association* 316 (1, July 5): 51–62.

Chavez, Luis O., Karen Torres, Maria Duarte, Salim Surani, Sharon Einav, and Joseph Varon. 2015. When Terminal Illness Is Worse than Death: A Multicenter Study of Healthcare Providers' Resuscitation Desires. *American Journal of Palliative Medicine* 148 (4, October): 771A.

Gawande, Atul. 2016. Quantity and Quality of Life: Duties of Care in Life-Limiting Illness. *Journal of the American Medical Association* 315 (3, January 19): 267–269.

Gould, Carolyn V., Richard Rothenberg, and James P. Steinberg. 2006. Antibiotic Resistance in Long-Term Acute Care Hospitals: The Perfect Storm. *Infection Control and Hospital Epidemiology* 27 (9, September): 920–925.

Halpern, Scott D. 2015. Toward Evidence-Based End-Of-Life Care. *New England Journal of Medicine* 373: 2001–2003.

Head, Barbara. 2011. Twilight Ethics: Dilemmas at the End of Life. *Topics in Geriatric Rehabilitation* 27 (1, January–March): 53–61.

Javier, Annie. 2016. Tattoos, Beer, and Bow Ties: The Limits of Professionalism in Medicine. *Journal of the American Medical Association (JAMA) Pediatrics* 170 (8, August): 731–732.

Levins, Hoag. 2016. End-of-Life Care Practices Not Based on Evidence. *Leonard Davis Institute of Health Economics*, June. Available at http://ldi.upenn.edu/news/end-life-care-practices-not--based-evidence (retrieved September 19, 2016).

Lindsay, Ronald A. 2009. Oregon's Experience: Evaluation the Record. *American Journal of Bioethics* 9 (3, February 26): 19–27.

Meier, Diane. 2016. *Medical Errors at the End-of-Life: Matching Care to Our Patients' Needs with Palliative Care*. Webinar, National Patient Safety Foundation (NPSF) Professional Learning Series, July 20.

Nuila, Ricardo. 2016. Whose Job Is It to Talk to Patients about Death? *Atlantic*, August 18. Available at www.theatlantic.com/health/archive/2015/08/palliative-care-medicare-end-of--life-ethics/400823/ (retrieved September 19, 2016).

Periyakoil, V. J.. 2015, November 18. Pitfalls for Proxies. *New York Times* (November 18, 2015).

Profeta, Louis M.. 2016. I Know You Love Me—Now Let Me Die. *Pulse LinkedIn*, January 16, 2016. A.vailable at www.linkedin.com/pulse/i-know-you-love-me-now-let-die-louis-m-profeta-md (retrieved September 19, 2016).

Span, Paula. 2016. The Prognosis Is Upsetting for the Doctor, Too. *New York Times* (July 5, 2016).

Volandes, Angelo E. 2015. *The Conversation: A Revolutionary Plan for End-of-Life Care*. New York: Bloomsbury.

Volandes, Angelo E., Michael K. Paasche-Orlow, Aretha Delight Davis, Robert Eubanks, Areej El-Jawahri, and Rae Seitz. 2016. Use of Video Decision Aids to Promote Advance Care Planning in Hilo, Hawai'i. *Journal of General Internal Medicine* 31 (9, September): 1035–1040.

Ethics End-of-Life Cases

Abstract

Nineteen cases of end-of-life patient stories are explored for common themes, issues, and problems. Clinicians can use these cases to extrapolate solutions in their day-to-day involvement with patients, using best-practice techniques. Based on the data explored in Chap. 2, these patients represent a broad spectrum of communication issues present between clinicians and their patients during end of life. As teachable moments for rising clinicians, there are questions following each of the cases for academicians to highlight in discussion.

Keywords

End of life cases • Case studies

During almost a decade and a half of work and learning in the field of ethics, I have seen a hybrid of issues that most probably represent a broad spectrum of communication dilemmas. I believe in the concept of case-by-case evaluation, which means: When you have seen one case, you have only seen one case. Each case is very different; a shift in one dynamic or demographic will change the outcome of that particular situation. I am, however, hopeful that these cases are representative of some obvious communication errors. Each analyzed case has been approved for use by the Institutional Review Board (IRB) in consideration of patient and surrogate privacy. Further, all names, ages, and other specific information have been adjusted with an effort to keep the message clear and to protect the patient. These cases represent the qualitative analysis of the study presented in Chap. 2, from years 2013 to 2016.

© The Author(s) 2017
K. Benton, *The Skill of End-of-Life Communication for Clinicians*,
SpringerBriefs in Ethics, DOI 10.1007/978-3-319-60444-2_4

4.1   Case 1: Dean

Dean was a sixty-one-year-old African American male who had been living at the nursing home with chronic obstructive pulmonary disease (COPD) and a recent lung cancer diagnosis. His chronic illness became much worse when he suffered the suicide of his son 3 years prior to his diagnosis. At that time, living independently, Dean stopped attending to his illness and became depressed. His only living next of kin was a sister-in-law who was intermittently involved with care and, ultimately, was the one to decide he required nursing home placement. She had a habit of disappearing from involvement for months at a time, refusing to return calls from the nursing home. Dean had refused all cancer treatment and made his wishes known for end of life multiple times in the past few years. Psychiatric care and counseling were pursued by the nursing home. He was placed on antidepressants as well. This helped his mood some; however, it did not change his wish for no further aggressive care, which allowed his disease process to continue. The patient self-consented to a DNR order, but hospice services were not discussed despite the patient's wishes. When the director of nursing (at the nursing home) was questioned about the patient's non-hospice status, she explained "that the discussion just never occurred; the nursing home is not reimbursed for skilled care when the patient is under hospice, so we don't pursue hospice unless specifically requested by the patient or family." It is difficult to understand why a request for no aggressive care does not align with a code or hospice discussion; however, this deficit in skill and communication is fairly typical in many nursing home environments.

When Dean's respiratory status worsened and infection set in, he was quickly transferred from the nursing home to the hospital; his DNR transferred with him. All other treatments were pursued, including placement on BiPap. No discussion of end of life was approached at admission, and he lost his ability to communicate before the end of the third day. With the sister-in-law unreachable, there was no advocate to consent to de-escalation of BiPap, medications, or hospice care. We began working on temporary guardianship of a local, neutral surrogate, but that was likely to take more time than Dean had. The patient was dying. He was stuck on a BiPap machine to maintain respiratory status and not intubate. A facial wound was developing. Despite Dean's status, hospice could not admit this patient without a signed consent. Ethics advocated that we uphold his stated wishes and the best interest of the patient by palliating the patient in the hospital. The attending physician was not willing to write comfort orders with a discharge of BiPap without Palliative Care involvement: "I can't make a comfort order! Why can't Palliative Services make the order?" Ethics relayed to the attending that too many days had been spent working toward hospice and now Dean's death was imminent. Time had run out for a Palliative Care consult. "Fine. I'll stop what I'm doing to do that!" Dean died alone, but in peace and under comfort measures in a matter of hours—but the unfortunate circumstances that unfolded prior to that very quick passing were anything but peaceful.

- Why do primary care physicians and specialists feel uncomfortable removing care and palliating death?
- Why are the obligations to dying patients placed last on the physician task list, when it might be argued they should be first?

4.2 Case 2: Abby

Abby was a fifty-five-year-old Caucasian female with end-stage renal disease. She was referred to the hospital for a neurosurgery evaluation due to intracerebral hemorrhage. She required intubation and eventually escalated to a need for trach and PEG. She was also documented as being noncompliant with dialysis, and she had no family or power of attorney (POA) living nearby. Abby was assessed as having full capacity for medical decisions once she improved somewhat neurologically and was stable on the trach. Ethics was called over concerns with compliance and futility of overall treatment. The patient was requesting full code and full aggressive care, though some days she would refuse dialysis. Physicians documented opinions that aggressive treatment was not likely to improve her prognosis due to years of drug use and noncompliance. Ethics then spoke with Abby, who insisted on full aggressive treatment despite concerns about compliance and the doctor's assessment that treatment would likely not improve her condition. She explained that she had made some poor decisions throughout her life but wanted to change and be a part of her family. We discussed the incongruent care decisions. Why did she refuse dialysis? She resented the question, stating she only refused infrequently when she "just felt too bad to make it through," but overall wished to live on artificial support and make amends with those she loved. She agreed to be compliant to be accepted into a vent facility out of state, closer to her family. She was able to be with children and grandchildren for 3 months and passed on the vent at the nursing home.

- Why does the connotation from noncompliance lead clinicians to immediately assume a negative behavior without investigating the reason for such behavior?
- Can the term "futility" be used with objective opinion when the reason for continued care is not understood?
- Can the choice for informed artificial support be a choice for quality of life?

4.3 Case 3: Melvin

Melvin was a seventy-seven-year-old Caucasian male who was admitted in February with sepsis, altered mental status, and respiratory failure. The patient had end-stage renal disease (ESRD) with hemodialysis three times a week. Melvin was a diabetic with severe peripheral vascular disease and dry gangrene. He had been advised to have a below-the-knee amputation to address these conditions. Melvin initially declined the surgery, stating he did not want to live without limbs, but later agreed when his family was convinced by physicians that prosthetics and therapy would offer him the quality of life he desired. He was eventually discharged but never strong enough to walk again.

The patient presented in the Emergency Department (ED) with altered mental status in May, again with sepsis, and respiratory failure. He also had a past medical history that included AMI (acute mesenteric ischemia or acute myocardial infarction), arthritis, coronary artery disease, ESRD, gastroesophageal reflux disease (GERD), hyperlipidemia, hypertension, seizures, a fractured hip from a fall, and left popliteal artery bypass. After being admitted from the ED, he continued to decline and received a PEG tube placement.

Despite Melvin's history of nine prior admissions within the past 6 months, as well as a prior amputation, seizure disorder, hypertension and diabetes, Ethics was not consulted until the twenty-fourth day after his May admission. At the time of consult, a second above-the-knee amputation was being considered. There had been no mention of less aggressive care, hospice, or involvement with palliative care. As written in prior admissions, Melvin had expressed that he did not want to live with any further amputations, and that he was having trouble accepting the first surgical amputation. However, the daughter had informed the physician that her dad had changed his mind and wanted the second leg amputated. The second procedure was done despite the back and forth on what he really wanted. Melvin never recovered mentally upon realizing what had been done to him.

When Ethics was called to recommend less aggressive measures, no physician had yet suggested a change in code status or a move toward hospice care. It is extremely rare for the family to bring up hospice unless led by the provider or care team. Ethics spoke briefly to the daughter about the septic condition and decreasing respiratory failure in terms of end of life. A family meeting was planned. The nurse called Ethics to inquire as to who informed the family that the patient was dying because he was "stable in the ICU." Ethics explained to the nurse the importance of being clear with the family on the low likelihood of recovery. Stability in the ICU with two lost limbs, sepsis, and respiratory failure was not a positive situation, and it should not be explained in such a way by the team that it would confuse the family.

When the family heard of the comfort option, they wished this had been presented before the first amputation. They had mistakenly thought that Melvin would have suffered more if they had not gone through with the procedures. The family made the patient DNR and comfort only; however, physicians felt end of life was imminent so no hospice was called. After four days of no change, and Melvin still in pain, his family began to feel uncomfortable with their decision. Without hospice to explain the end-of-life process, the family wondered if they should be doing more for him. A hospice consult was advised and counseling and pain management assisted with understanding. Melvin went home and lived three additional weeks with his family. Much of Melvin's agony could have been avoided had this same conversation occurred during any of his earlier nine prior admissions.

- Why does it feel more reasonable to escalate care in dying patients?
- Why are questions not asked regarding quality of life as a point of nursing assessment?

- Should a clear end-of-life plan or a hospice consult accompany a comfort shift for continued family support in a hospital?

4.4   Case 4: Lila

Lila was an independent and relatively healthy seventy-nine-year-old African American woman who lived alone. There was some concern by her children of the aging process, forgetfulness, fragility, and safety following a series of fairly minor strokes, but she was doing ok. One of the children was in healthcare and knew the importance of achieving a balance of life quality and quantity in the twilight years. The children checked on their mom every day and attempted to arrange transfer to assisted living as soon as she would agree; in the meantime, they tried to encourage more help at home, but she refused. One day, she experienced a catastrophic fall and was unable to call for help. There was an accidental mix-up between the children on whose day it was to call to check on mom. Each sibling thought the other was calling and therefore twenty-four hours passed before she was found. This riddled the devoted and well-meaning children with unnecessary guilt and anguish. Unfortunately, dehydration quickly led to kidney failure and then into sepsis. Because of the daughter who worked in healthcare, the family looked to her for decisions. Aware of their mom's stated wishes (though many children would instead have allowed their guilt to push for aggressive measures), they relayed, without prompting from any physician, that the patient would not want machines and would want natural death if prognosis was poor. The doctor simply said "Ok," with no questions or support, no offer for hospice, and no information to give them for what came next. Orders were written for DNR, withdrawal of BiPap, and the addition of morphine. Fifteen minutes later, the patient suddenly expired.

The family was shocked that they had consented to a withdrawal without fully understanding what that meant. They had imagined taking someone off life support was what withdrawal meant. They pictured days of hand holding before death. Instead, the family was left feeling guilty that they had agreed, in essence, to euthanasia, and they felt burned by the process.

- Did the doctor do anything procedurally incorrect?
- What could he have done to support the family?
- Why did the physician step out when no treatment orders were needed?
- Why are physicians excused from the explanation of how death occurs? Do you feel comfortable with that explanation?

4.5   Case 5: Ashley

Ashley was a thirty-three-year-old African American female with a rare aggressive breast cancer. Ashley had a husband and two small children and had already been through one surgery and subsequent chemo/radiation. She was admitted for a second procedure with a plan to be followed by additional chemo. During surgery, the surgeon was unable to remove the cancer and was surprised by how aggressive the cancer was. The patient was somewhat unstable following the procedure and was placed in intensive care. Two days post-op, an ethics consult was called by the nurse with concern that no prognostic information was being explained to the patient or spouse.

Ashley and her husband had asked multiple times what the outcome of surgery had been. Though the surgeon had very quickly explained, "The patient did fine," he did not explain the metastatic disease or the likelihood of no treatment plan after surgery. Instead, when the nurse called to ask that he come back to speak with family, the surgeon stated that he deferred to the oncologist. This was complicated by the fact that the oncologist was not seeing this patient since she was admitted to his care only for surgery. In order to maintain a direct chain of command with regard to health care, most oncologists see patients in the hospital when there is a diagnosis made in the hospital, when chemo is administered in the hospital, or when complications from oncology treatment cause the admission. All other concerns are referred back to the patient's attending physician. In this case, the attending physician would be the surgeon, until transfer to ICU, where it became the intensivist. The surgeon had replied to the nurse, "The oncologist can give more information when she is seen outpatient." In theory, this might be somewhat understandable since the physician is not an "expert" on the stage or prognosis of this very likely horrific outcome. However, he was the attending at this point, the captain of this admission. The patient and her family were cheated out of an explanation, and Ashley continued to deteriorate. The intensivist took over and still did not address the cancer prognosis. The patient's children could not visit because she was in intensive care, and sadly she worsened and became unconscious.

In this case, the focus remained on surgical complications instead of the cancer. This is an important mistake because a thirty-three-year-old with surgery complications should very appropriately be treated to the end; but a thirty-three-year-old with metastatic end-stage cancer with surgical complications may want at least the option of going home on hospice in her last alert days, to be with her spouse and children. That chance was never given.

Ethics explained the prognosis when everyone else refused to take ownership, after the patient had coded and when the decision to withdraw became the recommendation. Those children never saw their mother again.

- Why couldn't one person (a clinician) be charged to follow and gate-keep all information during a stay despite a transfer of specialties or units so that family and patient stay informed?

4.6   Case 6: Henry

The patient was a forty-year-old African American male. Henry was brought to the hospital by emergency medical service (EMS) after calling a loved one, who relayed that she believed he might have been having a stroke. The hospital determined that Henry had a brainstem hemorrhage. Ethics was called after a physician documented concern for the family members who would have to make tough choices for Henry because of his poor prognosis.

Ethics scheduled a meeting with the family to include daughter, father, step-mother, and siblings. The conversation included informing the family of how serious Henry's condition had become. The family explained that the physicians had not yet spoken with them about any prognosis. The daughter was not aware of how sick her father had become and was frustrated that other family members were keeping information from her and "treating her like a baby." Henry's siblings wanted to keep pursing aggressive treatment, but his daughter voiced that her father would not want to live with mechanical support. The rest of the family was willing to make Henry a DNR, move toward comfort care, and withdraw in support of the daughter's affirmation of her father's wishes. They did not want hospice or palliative care involved, so the attending wrote the orders. Once this was decided, no physician or team member met with, called, or discussed anything with the family again. The family needed support for their decision and instead communication ceased. It was if there were no longer a patient to treat. Four days after the initial meeting, the daughter had not officially set a time to start withdrawal. I called every day to check in with the daughter, but she stated, "some nurses had relayed they saw improvement."

When withdrawal is delayed, there will likely be an issue with final decisions. It is human nature to speak with other friends and family; although these individuals have no professional information on the patient's case, they want to add their two cents or share information on how their loved one miraculously recovered.

In Henry's case, some of the family members called Ethics and claimed that the daughter had psychiatric issues, and they had decided to fight her on the decision. It was true that the daughter had been seeing a psychiatrist for the past year. These family members wanted to overrule the daughter's decision on the basis that she was too young and mentally too unstable to make decisions. The daughter was the legal next of kin, so Ethics informed the family that the hospital does not make judgments on decision-maker capacity (only a court can do that) and that they would be upholding what the daughter decided.

After a conversation with the family, the daughter informed the hospital that she had changed her mind and wanted to try aggressive measures (trach and PEG). She felt too pressured by the family to be held responsible for her dad's death. She felt pursing a trach and PEG gave her more time to make decisions. Because she changed her mind, Ethics had to uphold her new decision and encourage the physician to try to continue to act in Henry's best interest. After the trach surgery, Henry required placement in a long-term care facility. It took nearly 3 months for the family to finally find a facility that would take him on vent. Henry was finally discharged to a rehab with trach and PEG, but was subsequently re-admitted three times within the next 3 months for seizures, infections, and PEG aspiration issues. The quality of life or lack thereof continued. This patient died of sepsis after a sixth admission.

• Why can't families be required to attend physician/team discussions to relay concerns on a bi-daily basis?

4.7   Case 7: Larry

Larry was a fifty-year-old Hispanic male with metastatic lung cancer who sought a second opinion outside of his hometown when he ended up in the Emergency Department with dyspnea, prior to a follow-up appointment with a new treatment center. Larry had previously been told by a local oncologist, per the notes, that he was terminal and there were no treatment options. His family seemed to be holding out hope that the new treatment center could save him. The patient informed the medical staff that he wished to remain full code but seemed to be in denial about the prognosis. There was no oncologist on the case during admission because Larry's original doctor had been fired and his new physician was out of town; thus, a hospitalist was managing the acute issue and consulted Ethics.

During the ethics meeting, the patient and spouse expressed distrust in healthcare in the hometown area. Both the patient and the spouse had been dismissed by the hometown hospital after presenting symptoms that eventually led to a cancer diagnosis. This dismissal led to the distrust even after the diagnosis was found to be positive for cancer. The patient and spouse were relying on false hope they had been given after the diagnosis they received at the new treatment center. Both were shocked to hear that the cancer was "metastatic" and "not curable." They claimed that although the local oncologist had not offered additional treatment, he did not inform them that cancer had spread, nor did he or the new treating physician use the word "incurable." The couple did not want any information from my center because of their earlier distrust combined with their refusal to hear anything negative.

Sometimes the extra mile is necessary for a patient who feels wronged. The human thing to do is to get answers they trust. Ethics tried to get a cancer treatment center representative on the phone for the family meeting, but no one was available. Ethics informed Larry and his wife that they should not "lose hope" but should recognize the limits of medicine and make choices in Larry's best interest (such as not burdening Larry with trips to the new treatment center and further aggressive treatment). Larry's wife completely shut down the code status conversation on the grounds that they had already heard the facts three times and discussing it would just upset Larry. Both were willing to discuss hospice; however, the wife claimed this was the first mention of hospice care/care at home.

As with many families, sometimes the "Do Not [Resuscitate]" conversation is more difficult than "Do [Give Quality Comfort Care]." The frame from which a decision is based can make all the difference.

After the meeting, Ethics planned to speak more about code status with the wife alone in order to respect her wishes to protect the patient from the conversation; the patient agreed such conversations were too upsetting for him. After the initial meeting, the patient's condition worsened, making travel impossible; Larry continued to let his wife speak on his behalf, making coming to terms with his prognosis very difficult. The family continued to hold out hope that the new cancer treatment center would offer some lifesaving treatment. After hearing from the cancer treatment center physician that they would be unwilling to treat Larry again due to his worsening condition and terminal illness, Ethics asked that the physician be conferenced in and relay that message himself. The physician reluctantly agreed. Larry and his spouse then agreed to hospice. Larry was discharged with hospice care and agreed to Allow Natural Death, a more acceptable language than Do Not Resuscitate.

- How do you respond when a spouse or surrogate dismisses an end-of-life discussion (in a conscious and alert patient) out of feelings of discomfort or protection?

4.8 Case 8: Sara

Sara is an eighty-two-year-old Caucasian female who presented to the Emergency Department due to difficulty with speech. Her family noticed the problem and brought her to the hospital after the speech issue persisted for five days. A CT scan in the ED revealed a frontal lobe brain tumor. The physician spoke with the family about the CT findings and treatment plan. The family did not want to notify the patient of her diagnosis. The physician asked that they please speak with the patient in the morning.

Why do we allow the family to dictate care? Sara should decide whether she would like to hear her diagnosis. Ethics was called after the family refused to notify the patient of her diagnosis. Unfortunately she became incapacitated due to symptoms. The family was unsure how to proceed with the treatment plan. Two children wanted a biopsy and aggressive care, while two children wished for hospice. There was no guidance given to the family, only a menu of options initially given by the hospitalist.

Ethics had an initial meeting with the children, who repeatedly deviated from discussing the patient's wishes to arguing with each other over family issues. Ethics then spoke with Sara after she became more alert. Her family was asked to stay out of the room in order to not disrupt or pressure Sara. She had a difficult time processing the information about her diagnosis and wanted to "think about it" when a brain biopsy was mentioned. However, she did not hesitate to inform Ethics that a grandson should be her legal decision maker, and she signed a power of attorney, stating, "My kids fight too much." The children were all informed of her decision. They all seemed ok with the grandson as POA.

Ten days later, Ethics was called back by the physician because the family was still unresolved in the patient's plan-of-care decisions. The grandson did not want to consent to the brain biopsy or continue with radiation/chemo to treat the tumor. Other children disagreed. The grandson was unconvinced of the right choice since consulting physicians all had different opinions:

- Hospitalist 1—recommended biopsy "just to know"
- Neurology—recommended steroids, repeat MRIs, then yes to biopsy
- Neurosurgery—too risky to have biopsy for someone this age when there is little that can be done
- Hospitalist 1—"Maybe you should get another neurosurgery opinion."
- Hospitalist 2—"I wouldn't do this to my dad or granddad .... what do y'all expect?"

Different doctors gave the family varying opinions instead of all coming together and collaborating on a cohesive prognosis. Ethics spoke again with the POA and felt that hospice was the best choice based on Sara's wishes as relayed by POA. Because of multiple medical opinions, POA wanted to be guided by the neurosurgeon and focus on making choices for the best quality of life for his grandmother. After the family backed down on the biopsy decision, the MD suggested that Sara be discharged home with medication. Hospice agreed to meet with the family in their home and make decisions on how to best care for Sara at home. The MD informed the family that the patient could always return to the hospital if they want to have the biopsy performed. In the meantime, Sara could be comfortable at home.

Much of the time, goals can be easily set once ego takes a backseat to teamwork.

- When should a team meeting for differing opinions be an absolute and how can you facilitate this?

4.9   Case 9: Martin

Martin is a ninety-year-old African American male who suffered a stroke that wiped out his entire cerebellum function. Immediately after his evaluation, the doctor walked into the waiting room and made what the family called a "candid" speech, for which they were so grateful.

"I'm not going to sugarcoat it, he has had a massive stroke," said the physician, "but it is not completely hopeless." That last line grabbed family's attention. They heard a well-meaning, honest doctor who had hope for survival.

The daughter replied, "That is all I had to hear. If it is not hopeless, we will do everything." The family was extremely religious; therefore, to deviate from hope by "giving up" on artificial mechanisms was to give up on God, and some alignment for withdrawal was seen as an active form of euthanasia. The family did not waver or wish to hear any negative information from then on. "God works miracles" became the mantra.

By the time Ethics was called in, the family was considering a trach and PEG. Knowing faith was important to this family, Ethics did not focus on autonomy/patient wishes or prognosis, Acknowledging the family's religious beliefs, the Ethics consultant also chose religion to guide the communication. The representative from Ethics told them that he was a person of faith and believed in miracles and that the choice they had been given was only from a scientific point of view. Ethics relayed the good news that God heals despite machines, but reminded the family that, in fact, not one miracle in the Bible called for life support or dialysis. Hope should be met where it is, realized and acknowledged.

Ethics also explained suffering and the good news of eternity for those who believe. But, in addition, if Jesus were to call Martin home right that moment, as a full code, hospital personnel would have to use their own oaths as healers and attempt to shock him back to this world. The family quickly changed the patient's status to Allow a Natural Death. They were uncertain about withdrawal but did know for sure they would "not want him cut on anymore; that was not natural." The removal of the trach/PEG also entered into the conversation.

Next, Ethics talked about responsibility—whose responsibility it was. Not theirs. They were not deciding whether the patient lived or died. A massive stroke had decided this situation; the family needed to examine how end of life might look. But even as the patient was intubated, the family made contradictory comments indicating they were not being fully informed: "Yes, Dad did say he never wanted machines." "We just felt it would make us responsible for his death if we say pull the plug, and I cannot live with that on my conscience. I have a soul to protect for eternity." When responsibility was removed, the daughter, wrought with concern that her father's life rested with her decision, literally breathed a sigh of relief. Might he live a bit longer intubated? Yes, but the family would not move forward with trach, and intubation cannot hold a patient here forever. Would Martin live with fewer complications? No. Did he want any of this? Ethics soon learned that this was a man who had never gone to the doctor a day in his life. He hated hospitals and doctor's offices, but his last 2 weeks of life were not spent on his beloved farm, where he had resided for sixty-one years. Instead, they were spent in an environment he detested. After considering all options, with two more conversations covering a series of questions, withdrawal took place three days later.

- Is suggesting making a patient's church leader be a part of the family's conversation a good option?
- What does the role of a higher power play in the work of a clinician, specific to communication?
- Without a common background, can a clinician who is of a different faith or no faith at all communicate as clearly with someone of a different faith?

4.10   Case 10: Michael

Michael was an eighty-three-year-old Caucasian male who had suffered a stroke several years earlier. His stroke had left Michael in a vegetative state and vent dependent. He had no designated power of attorney (POA) or living spouse at the time of the stroke, leaving his four sons as joint decision makers. Two of the sons felt strongly "not to give up" on their dad. The caretaker son, James, was the most informed of all the children because he was the usual contact person for the staff/physicians, since he was always in the room. Michael had been readmitted to the hospital every few months due to infection, vent/debilitation, and accidental harm by family when a mucus plug was missed or dialysis did not drain correctly. Ethics was consulted each of the previous three times the patient had been readmitted due to staff distress from having to continue treating the patient and discharging him back home. The caretaker son was willing to consider boundary setting and DNR after his father had been vent dependent for years and on his thirteenth hospitalization at the time of the ethics consult. Due to disagreement between the sons as joint decision makers, the patient remained full code and the sons remained unwilling to de-escalate care. The sons felt that Michael's wishes were being respected by a life on vent/PEG, but a decision about boundary setting had never been made by the patient. The decision was left to the interpretation of the sons. The decision makers truly felt their father's will was being respected.

- Can a family responsible for the patient make withdrawal decisions when they truly do not know the patient wishes?
- Is it more fruitful to call together the entire family for thoughtful discussion when all are advising decisions?
- Do we inexcusably judge families for their caretaking and choices in patients we cannot heal without leaving room to process the distress they may be experiencing to choose when to allow natural death?
- Is the responsibility of artificial machine/prolonged caretaking too much on a family?

4.11 Case 11: Tom

Tom is a forty-two-year-old Caucasian male with a history of severe mental retardation and had previously undergone a bilateral enucleation. The surgery was performed due to the patient's mental status and his inability to refrain from picking at his eyes. His past medical history also included dementia, chronic constipation with megacolon, GERD, hypertension, anxiety, esophageal dysmotility, osteopenia, vitamin D deficiency, diverticulosis, impulsive control disorder, and pica. The patient was taken to a doctor's office by his full-time caretakers from the personal care home in which he resided, and they claimed that the patient had been having chronic constipation. The caretakers had been working with gastroenterology to fix the constipation by having Tom drink a GoLYTELY prep. On the morning of the doctor's visit, Tom had vomited and presented with a somewhat distended abdomen. The gastroenterologist directed them to the ED, in which the caretakers gave a detailed medical history of the patient, as well as constipation concerns.

The ED physician reported that Tom was actually having bowel movements (up to five per day). The MD suggested to the family that she would be willing to perform a manual disimpaction and send Tom home since his x-ray did not show his chronic megacolon. The caretakers wanted Tom admitted anyway because that had been the original recommendation at the gastroenterologist's office. Tom was admitted to the hospital per the wishes of the caretakers and was evaluated by gastroenterology. Upon admittance, the medical assessment concluded that Tom needed a colectomy in order to resolve ongoing gastrointestinal issues; however, the family, who was contacted subsequent to admission, refused the treatment plan. The physician was concerned about the continued use of GoLYTELY due to the Tom's para-esophageal hernia and the associated risk of aspiration. Ideally, Tom needed the hernia surgically corrected and a total colectomy.

Ethics was called by the social worker due to the family's refusal to have Tom undergo the procedure and their refusal to take him back home. Ethics communicated with the home care director about the proposed care. The director also informed Ethics that Tom did not have a one-on-one caretaker who could intervene if he was attempting to pull the plug on his colectomy bag, which was likely. Ethics then spoke with Tom's family, who felt that the likelihood of his constipation issues returning were slim. The patient's nieces, who were next of kin thought that the hospital had likely "cleaned him out," where they had been unable to do so at home in the past. Now that Tom was clear, his problems would go away—making further surgery unnecessary. The nieces were both concerned about Tom's quality of life "with a bag." He already had a history of being a "picker," and they felt a colostomy bag would prove to be a problem. After getting their questions answered during the ethics conversation, the nieces agreed to make a final decision the next day.

The hospital staff was only viewing the need for next steps instead of considering patient behavior, which might well hinder his quality of life and his admittance in an excellent personal care home, which had been difficult for family to find. Ultimately, they refused the hernia and colostomy surgery but stated that if issues did not improve, they would consider hospice.

In days, Tom was eating normally and having regular bowel movements. The nieces agreed to allow Tom's discharge back home a week after his initial admission.

- Is quality of life a concept to truly outweigh longevity?
- What are the most important ethical points of consideration when the patient does not have capacity to understand when quality of life versus quantity of life is being evaluated?

4.12 Case 12: Rosa

A seventy-two-year-old Asian female was transferred from an outlying hospital to a more acute facility and then discharged to a long-term acute care (LTAC) facility. Rosa arrived with acute respiratory failure attributed to community-acquired pneumonia and COPD. She required intubation and mechanical support from a ventilator just prior to discharge from the first facility. After arriving at the hospital, she developed atrial fibrillation with rapid ventricular response, thrombocytopenia, and hospital-acquired pneumonia. In addition, the patient developed an acute kidney injury (which resolved), anemia, and a Stevens-Johnson reaction, which also improved. Moreover, the patient had stage IV breast cancer that had been surgically removed with prior chemotherapy and radiation therapy. Finally, there was a past history of myocardial infarction and COPD with continued tobacco use, despite an MD's suggestion that she quit smoking.

Rosa and her family had been followed by Palliative Care during her earlier admission. Palliative Care had introduced hospice, but the family had been unwilling to pursue that course of treatment and chose to send the patient to the LTAC. Weaning proved unsuccessful, and Rosa had to be moved to a more permanent facility on the vent if she wanted to continue with mechanical ventilation. Ethics was called to speak with Rosa and her husband.

The patient chose to continue with placement in a vent facility, but her expressed wishes in the past had been for comfort care only if her prognosis was poor. The concern was that Rosa had not been fully informed that she was at her end of life, even though Palliative Care was very candid with the spouse. Much of the language in consultation with patient included such questions as "Do you want to continue to try to get better?" That was an unfair and inaccurate question since that could no longer be the goal. Ethics spoke with Rosa, who stated that she believed a vent facility was a place she could go to improve. She was overwhelmed and scared by the decision to withdraw or continue with placement at another long-term facility. The patient decided to place with hospice after being fully informed of the poor prognosis and end-of-life care issues. She survived withdrawal to die in peace in a better environment.

- How comfortable are clinicians talking with respiratory suppressed (vent and trach patients) about end of life?
- Is there a reasonable element of protecting the patient against a withdrawal discussion when they are alert?

4.13   Case 13: John (called Jack by family)

Jack was a sixty-two-year-old independent Caucasian male who developed chest pain while at home working in his yard. EMS brought him to the ED. The patient was admitted after having an abnormal electrocardiogram, slight elevation of troponin, and higher than normal total creatine kinase (CK). After his admission, Jack also developed supraventricular tachycardia and hypertension with a history of diabetes and hypertension. Jack's case was determined to be acute, and he received immediate catheterization after his admission from the ED. The catheterization procedure revealed obstructed coronaries; therefore, he was taken in for emergent coronary artery bypass grafting (CABG). The CABG procedure went without complications, but Jack's condition worsened postsurgery. The patient was hypotensive (requiring multiple pressors and an aortic balloon pump), experienced respiratory failure, and became hypoxic. Jack was intubated and placed on a ventilator. After the coronary care unit (CCU) stabilized the patient, he was treated for cardiogenic shock and pneumonia. While in the CCU, Jack developed thrombocytopenia, renal failure, and liver failure. As Jack's hospitalization time increased and his health status did not improve, Jack's wife expressed her wish to withdraw him from the machines.

Ethics was called to speak with the wife because the MD wanted to continue aggressively treating Jack in hopes that he would improve. The physician continued to ignore the wife's request, and stated, "Why would you have called EMS if you did not want to continue?" "You are an idiot and should allow me to help your husband." The physician never listened to the wife's response. Ethics spoke with the wife, who explained that Jack had made his wishes known after losing a son. Both parents witnessed their son die after being placed on life support, and Jack had told his wife that he would never want to end up in the same position. His wife knew that he lived a very active lifestyle and would not be ok with the diminished quality of life after recovering from surgery.

Ethics discussed how quality of life can change as a person adapts to changes. Maybe her husband would be able to accept a less-active lifestyle for a longer life span? The wife remained convinced that her husband would want to be withdrawn despite the MD's opinion that he might benefit from further treatment. Jack's wife explained that she originally pursued aggressive treatment because she was confused and overwhelmed, not because it was in line with her husband's wishes. She called EMS so he would not suffer through the event and felt coerced to this point because little to no conversation had occurred before he was "all of a sudden on life support." She wished her husband's condition had been caught earlier, so he could have made some choices about treatment. Considering the current situation, she felt it was right to withdraw him and get him off life support.

In concert with her decisions, Jack was made a DNR, given comfort measures, and withdrawn. Ethics was called back in when Jack was sustaining off the vent and his family wanted hospice. The MD was refusing hospice orders and encouraging tube feeds to see how Jack would progress, despite concerns that he might have suffered anoxic brain damage and would need amputations in both hands due to ischemia.

Ethics spoke with the wife again, who agreed to continue with nutrition as long as Jack was comfortable. The plan was to wait for the patient to improve in order to allow him to make his own decisions. If he did not improve, he was still listed as a DNR and would not go back on life support. The wife remained adamant that she wanted hospice.

Ethics was faced with the challenge to find a consulting physician to write hospice orders since the attending refused. Jack was eventually allowed to be placed under hospice care and expired peacefully.

- How comfortable are we in healthcare with families who actually draw their own boundaries?
- Do you believe following a surgery the physicians and staff may push for full care simply to improve mortality rates following surgery, and possibly discount quality of life following surgery?
- Do clinicians explain the bargains and compromises patients may have to make specific to their quality of life, simply to stay alive?

4.14 Case 14: Maria

Maria was an eighty-year-old Hispanic female previously admitted to the hospital but then discharged to a nursing home 2 months later. During her previous admission, Maria had been suffering from a GI bleed associated with blood loss anemia, complicated by severe cardiomyopathy, advanced dementia, and diabetes. At that time, the hospital tried to contact family members to help make a plan-of-care decision for the patient but was unable to reach any family members. The Ethics Subcommittee decided that the best course of action for the patient was to make her a DNR and pursue a less-aggressive plan of care. Most DNR laws allow two physicians to concur with Ethics' support when there is no patient or surrogate to consent to a DNR. The patient was sent to the nursing home with a POLST (Physician Order for Life-Sustaining Treatment) form to prevent any future inappropriate aggressive treatment or intubation. The patient was discharged only when she became stable.

The facility that Maria was sent to had the responsibility to keep the patient under comfort care; however, while at the nursing home, Maria was found unresponsive with low blood sugar. Maria was brought back to the hospital for further treatment. Ethics called the nursing home to ask why the patient had been re-admitted when she had a POLST form. Ethics also asked why glucose labs and other basic care, including palliative measures, were not being given to the patient. The nursing home administrator was unaware that the patient had been re-admitted and was unwilling to work with Ethics on the case. The director of nursing did relay that their physician refused to have any end-of-life conversations with their residents and was unfamiliar with the POLST form. When Ethics asked how POLSTs were completed at the nursing home, they explained that Admissions staff checked the box and forms were presigned by the physician. The nursing home did have the phone number of a cousin, which they had not provided on the previous admission. Ethics reached out to the patient's cousin, who was willing to sign off on a new nursing home. The cousin had not visited regularly and was unaware how sick the patient had become, but she did know that the patient's wishes aligned with comfort only. The patient was discharged to a different nursing home with the same POLST and DNR, with the hopes that the new facility would better communicate and respect her wishes, ensure Maria's best interest, and allow her to remain in the facility under comfort care. The patient was placed under hospice care at the new nursing home.

- As the aged population increases, what responsibility do clinicians have to ensure a nursing home addresses end of life?
- Should hospitals and nursing home surveyors (regulatory agencies) be allowed to ding a nursing home for not addressing palliative care seriously and correctly?

4.15   Case 15: Randy

Randy was a twenty-nine-year-old Caucasian patient who suffered a severe neurological injury during a motor vehicle accident caused by overdose and cardiac arrest, resulting in anoxic injury during the accident. Randy underwent many surgeries to try to repair some of the damage, but ultimately, he became a severe quadriplegic with very limited neurological function (close to a vegetative state). He stayed on a trach and was tube fed. Randy had been admitted for infection from wounds and had suffered a hip dislocation during his stay at the hospital by the time Ethics was called. His father was demanding surgery, while three surgeons had consulted and refused the surgery due to their inability to perform surgery with successful ambulation and minimal risk. Dad felt it would improve Randy's quality of life.

Ethics was called by the physician to address goals of care and for discharge planning. Ethics spoke with the very frustrated family members. The family had previously appeared unwilling to cooperate in discharge planning. During the initial ethics conversation, Ethics discussed the complexity of the situation and the need to find another nursing home that would be willing to accept Randy since his dad did not want to take him back home at that time. He felt his son needed too much care and he, the father, needed to go back to work. The family needed an advocate to help manage their issues of distrust and frustration.

In the role of a healthcare professional who was looking out for patient/family's best interest, Ethics was able to establish trust and rapport with the family. Over the weeks of trying to find placement for Randy, the family continued to express distrust in treatment instead of recognizing the severity of his condition as the root of the problem when it appeared that the patient was digressing. Issues with his kidney function became a topic of discussion. Focused on Randy's need for hip surgery, the family had a hard time understanding why Randy's kidneys were a problem. They demanded better communication. It was discovered that although staff had been heavily involved and was extremely burnt out from this case, no physician (including the three orthopedists) had actually spoken with family about their issues with surgery. All physicians involved had legitimate reasons for weighing the risks and benefits as they had, but again, the actual communication was lacking. At the end of the 3-month search for a facility that would accept the patient, Randy coded and was unable to be resuscitated. He expired in the hospital after being unable to transfer to another facility.

- Should dedicated and regulated, timed phone calls be required of physicians to explain care and progression to family?
- Should communication with fellow specialists in a patient's care be required to attend an in-person meeting?

4.16   Case 16: Nina

Nina is a thirty-two-year-old African American female with sickle cell disease. The patient presented almost every single day in the ED asking for pain management during her sickle cell crisis, and eventually Ethics was called. Nina has a past medical history of deep vein thrombosis (DVT) and pulmonary embolism (PE), along with her sickle cell disease. Nina has also been diagnosed with schizophrenia, was a reported drug user, and had recently been released from jail. In the ED, medical staff had difficulties dealing with the patient due to her mental health issues, noncompliance, and lack of trust toward hospital personnel.

Ethics was asked to speak with the patient and address issues of care management, leading to chronic re-admission and staff moral distress when treating the patient. During the time of the initial consult, the patient left the admissions area before Ethics was able to speak with her.

Nina's case was followed on an outpatient basis in order to address issues and limited re-admissions. During the outpatient follow-up, an interdisciplinary team (Ethics, Pastoral Care, and a Palliative Care physician) met with the patient and her mother. The team learned that the patient had been fired from several clinics in the past, so she was lacking a primary care physician, a hematologist, and other supporting healthcare specialists. At the meeting, the patient's mother detailed a long history of discrimination for her daughter due to the fact that her illness is "not visible from the outside." Nina and her mother had developed a pattern of distrust for medical staff as a result of this discrimination. The meeting also addressed counseling or psychiatric resources for the patient to deal with her mental health issues. Nina's boyfriend had recently died, which seemed to be particularly difficult for the patient to cope with. She initially refused to meet with a counselor but later agreed to try it out. When asked about her goals for the future, the patient was silent. She did not see a future direction for her life beyond visiting the ED every day.

The team that met with Nina suggested arranging a volunteer position for her in order to give her some structure and meaning to her life. After the meeting, Ethics made an effort to reestablish the patient with a primary care physician. The selected physician was one for whom the patient had expressed some trust and positive feelings toward; the physician agreed to take on the patient, pending an agreement toward treatment compliance. A Palliative Care team agreed to work with the patient on pain management, advance directive planning, mental health issues, treatment goals, and personal goals. In addition, the Palliative Care team was hoping to connect the patient to a Medicaid program that would provide the health care she needs, to include dental care, treatment for other pain/injuries not directly related to sickle cell, and possible home health for the future.

- What are our ethical obligations to ongoing health and chronic disease issues in the Emergency Department?
- Should we be mandated to address social determinants along with care in a repeat ED user like Nina?

4.17   Case 17: Evelyn

Evelyn was a fifty-two-year-old Hispanic female with no known medical history. She was being seen for osteoporosis and receiving injections to help with that deterioration. You will note my personal adjectives to describe Evelyn, as she was also a close friend. She was an intelligent nurse practitioner, a mother of two adult children, and had been married for 30 years. When Evelyn contacted the physician with constant vomiting, she assumed it was from her injections. She had some weight loss during the previous 6 months accompanied by back pain, but as a medical professional, she was confident that it was nothing serious and was pushing through the symptoms. The vomiting worsened, which lead to dehydration. Evelyn was eventually admitted to the hospital. Shortly after admission, she suffered what appeared to be a pin stroke, affecting her neurological status. She was responsive and alert but clearly not at baseline, refused to eat, and had difficulty communicating. Evelyn was no longer able to make decisions. Physicians immediately began a work-up and ultimately found a small stroke (possibly from a medication she had taken), which they felt she would recover from. As she began to rapidly worsen, the physician decided to do a full CT scan. It was at this point that he discovered that Evelyn had very advanced non-small-cell carcinoma, commonly known as the nonsmoker's cancer. Her father had this same history and Evelyn's had progressed aggressively, causing clotting and secondary strokes. The prognosis was not good. The family was notified of the results. The physician was a family friend and quickly called in Oncology, pushing straight into fight mode with no consideration or mention of doing nothing. This patient was young, her case was unexpected, and doing nothing to treat the disease felt wrong. Oncology recommended what they termed on the chart "palliative chemo" but told the family simply, "We will begin treating in the hospital." Evelyn received one round of chemo and was somewhat stable—essentially as stable as she was going to be when she was discharged.

Family members described their experience in this way: "We were told so many different things from different physicians that we truly needed 'a team captain' who could somehow culminate all the medical plans from each physician and give us a clear concise picture of how we needed to move forward. There was massive confusion and misunderstandings as we tried by ourselves to understand all their quick and complicated medical jargon. You [Ethics personnel] were the saving grace that served as a 'translator' to guide us through the swamp of various diagnoses and medical plans." Instructions were given to care for Evelyn at home. Home health was set up and a plan for continued outpatient chemotherapy was developed.

Evelyn was never aware enough to understand her illness or prognosis. This brilliant, loving, and assertive woman no longer made any decisions. She moaned in bed and was only quieted with song. At home, she was often agitated and in pain, very aggressive with family caretakers, and almost in need of home restraints to be cared for (though the family did not use restraints). The family was beside themselves. They did not know what the future held, how long they should treat, what the goals were, or where to go next.

Ethics provided them with some counseling and the logistics of the case was discussed, void of emotion, to find what might seem like the best care. Ethics, in fact, provided them with the brilliant book by Ira Byock, MD, *The Best Care Possible*, and encouraged that they consider loving options. Transport to chemotherapy was not even a reasonable goal at this point. Evelyn could not be transported without restraints, and care at home was exhausting and endless. Ethics encouraged the family to make a calendar and ask for help from church, friends, and family caregivers; however, this was a mere Band-Aid to the lingering question of what was best for Evelyn.

Ethics then suggested that Evelyn not go for her next dose of chemo, and instead that Ethics go with them to her next appointment with the oncologist to meet with the consulting expert and ensure that the family understood all the recommendations. A family does not want to be the one to say "do nothing" for the young person they love so desperately unless that is the only option, and it was not in this case. The oncologist listened while the family explained the patient's agitation and pain, and the unrealistic nature of receiving treatment. The family also explained that they wanted her to live if possible. The oncologist said, "Well, she can certainly have chemo, but remember, it is palliative."

Ethics asked that the oncologist explain what "palliative" meant. The oncologist stated that "It will not cure the disease but may slow it down." Again, Ethics pressed the issue and encouraged the doctor to admit that the cancer had no cure and the treatment was palliative. The stroke and neurological damage was also irreversible. In essence, therefore, Evelyn's personhood at that point was permanent. "This is her new quality of life, correct?" The oncologist said, "Yes. That will not improve, only the cancer may slow down." Again, Ethics pressed, "They need to hear from you that a choice for no treatment and toward hospice is ok—something you, as a physician, would agree with." The doctor then said, "Yes, that would be very appropriate, knowing she will not get better than this, and ultimately does have an incurable, terminal illness."

Finally, hospice could be called into the home. Eventually, Evelyn required care in a facility to control the agitation and pain. She died very peacefully several days after her transfer into hospice.

- What prevents use of the term "dying" in a dwindling patient?
- Why is compassionate chemo considered compassionate, and should the oncologist feel obligated to offer it?
- Is palliative chemo or radiation appropriate in a patient with no capacity or quality of life based on surrogate definition?

4.18   Case 18: Ben

Ben was a seventy-five-year-old Caucasian male with rapidly advancing dementia and chronic obstructive pulmonary disease, requiring BiPap at night and oxygen during the day. Ben had been placed in a nursing home by a cousin (his only next of kin) months earlier, when he had no longer been safe at home. Ben needed the PEG because he stopped eating, and the nursing home demanded that he have it. His cognitive function came and went as it does with many patients with dementia. When he entered the nursing home, he had an advance directive, giving decision-making authority to his cousin and expressing his wishes for a natural death. This document was completed 5 years prior to his diagnosis for dementia, so there was no question as to his mental capacity to make such a decision. The nursing home social worker and nurses engaged in conversation with him on resuscitation wishes, where he very definitively changed his code status to a DNR, stating he never wanted tubes and machines.

On some days at the nursing home, Ben was very capable of holding conversations; he was able to make requests, and at times, he was even capable of meeting his own needs. On other days, he seemed catatonic. Over the course of a month, Ben began to stop eating much at all. Weight requirements at the nursing home required that the patient meet baseline or close to it. This required that Ben be transferred from the nursing home to the hospital for "failure to thrive," and possible PEG placement. The pulmonologist worked the patient up and quickly consulted surgery for the PEG. No staff or physician attempted to discuss or engage with the patient, calling the POA directly for consent. When Ben went down to surgery, the anesthesiologist chose to discuss anesthesia with him, since he "seemed to have clear capacity." Ben quickly refused the entire procedure, stating he never wanted tubes. He would not be tube-fed and would not sign consent. Anesthesia notified the attending, who stated, "He has dementia; the cousin signed consent; it is fine to proceed." Ethics was called.

Capacity is task-oriented. Ben had never been declared incompetent by courts and still had rights. Ethics re-asked the question and the patient again refused the PEG. Ethics had a lengthy discussion with the attending. Ethics told her that Ben did not agree with PEG placement. It was not a means to an end but rather a choice on change in quality of life and an end in and of itself. It was permanent. What happens when he pulls it? Will restraints be next? Dementia still left this man with rights. Ethics suggested we explain the situation to the POA and ask for her support in supporting Ben. Without the PEG and with malnutrition, he was hospice appropriate. Luckily, the POA also agreed to uphold the patient's wishes, understanding that clinicians do not ignore or force a patient, and frankly, a decrease in appetite was a natural way to die—as the patient wished.

Ben was not depressed. He had two advanced and terminal conditions and did not want to add artificial support to his regimen. He was sent back to the nursing home under hospice, with no plan to return to the hospital. His pain and symptoms remained controlled. He ate as he wished and lived an additional 7 months—thinner, happy, and autonomous.

- Is a PEG ever appropriate in a demented patient, and if yes how many times should it be replaced when they pull it?
- If a patient with partial capacity refuses a life-prolonging treatment, what is the obligation to the autonomy?

4.19   Case 19: George

George was a sixty-seven-year-old Caucasian male with a history of chronic diabetes and kidney disease. He was receiving hemodialysis three times a week and had been admitted to the hospital through the Emergency Department for pneumonia. Ethics was called by the case manager and the doctor regarding concerns over the patient's wife as the decision maker. She was rarely at George's bedside, but she was relaying patient's stated wishes for full code and aggressive care. George was intubated and re-intubated before Ethics could schedule a meeting with the wife (due to her long absences from the bedside). The wife considered DNR but did not feel it was appropriate based on George's stated wishes for aggressive care. Due to the wife's feelings as a decision maker and the patient's stated wishes as relayed by the wife, the patient remained a full code and continued vent with eventual need for trach and PEG. Eventually the patient become conscious and was able to participate in decisions. He coded three times during dialysis but was always resuscitated. Multiple staff, palliative, and ethics conversations with his wife present proved to staff that resuscitation was always his wish. He understood he might die during resuscitation but continued to keep these as his wishes, stating he wanted to fight until the end.

- Does it make clinicians feel unreasonably uncomfortable when a clearly informed patient decides against medical advice?
- Does moral distress from providing death/life decisions to prolong artificial support result in pressure to change a code status?

These cases represent only a snippet of the dilemmas that occur. The unfortunate reality is that they are neither rare nor unique. Such cases as these happen in day-to-day care in hospitals all across the nation. One challenge in authoring this book was to pick those cases that best represented the "usual" dilemmas. Now that you have read and discussed many of these cases, the question becomes, how do you skill your way out of needing Ethics to come in? How do we do this better for the clinicians, for the patients, and for the loved ones involved?

Reference

Byock, Ira. 2012. *The Best Care Possible: A Physician's Quest to Transform Care through the End of Life*. New York: Penguin.

The Skills of Communicating Clearly

5

Abstract

Although communication skills are often overlooked in clinical training, they are vital to the development of a good clinician. Since most of us speak from childhood, we assume that speaking is an innate skill, but communication is not the same as simply talking. As clinicians, all must learn how to talk to patients as well as our fellow professionals in ways that ensure the message we send is clear, coherent, and heard. This is a skill, not unlike being a good diagnostician or surgeon. It is precise and intuitive, and requires practice like any other art form.

Keywords

Communication • End-of-life communication • Physician communication • Advance care plan

Following the introduction to this chapter, the defined skills are divided into subsets. Some are followed with brief case examples for explanation, others by a quick list of items you can do or to use as a reference when you need quick advice on how to handle your patients' "elephant in the room" information.

5.1 Clinical Skills and Communication

All clinicians should be valued for their selection of a profession where most customers—that is, their patients—are having one of the worst days of their lives, not their best. Clinicians do not serve cocktails or food. They do not sell retail or work in hospitality or event planning. The discipline is riddled with anxiety and sadness, and therefore it becomes more important than ever to know how to communicate, empathize, relate, and listen. This book does not intend to criticize or to take away from the brilliance and intellect that must come with being a clinician. However, in a review of the stereotypical clinician and understanding why their communication

© The Author(s) 2017
K. Benton, *The Skill of End-of-Life Communication for Clinicians*,
SpringerBriefs in Ethics, DOI 10.1007/978-3-319-60444-2_5

skills often lag, it seems skewed that one can operate and reset a dead heart, replace kidneys, remove cancerous growths, put a tube down a person's throat for breath, or connect an intravenous line in for healing—but then despair at talking about the big picture of complex care. Communication is a skill that happens to require a different set of competencies, and its proficient use remains vastly undertrained.

Clinicians—perhaps because they view death as a failure of their trade—appear to be threatened by the end stage in medical care, so they are drawn toward abandoning the patient when the treatment they offer can no longer cure. This abandonment can usually be seen as cowering when considering talking about the end, quietly signing off when they can offer no more. Instead, an open and empathic conversation about end-of-life issues can be of much more use to a patient once all other options have been eliminated.

One of the key recommendations by the Institute of Medicine (and therefore even pushed federally) was the development of the skill of talking to patients during this time (2014). At the state level, mandates, guidelines, and regulations that encourage such communication skills would make a big impact and yet be both reasonable and fiscally conservative. Although the mere act of talking seems so elementary and has thus always been overlooked as a procedural concept needing knowledge, it is instead the one skill that sets apart the merely competent medical professionals from those who excel at their field.

Patient-centered care is a healthcare buzz word, but the art of kindness and communication should be at the heart of this care. Where are the hugs, the touches, the dialogue, and the conversations when time grows limited and treatment becomes purposeless? Discussions have all been skewed by structures set in motion that place documentation and a problem model for patients as one that overshadows basic touch, true empathy, value in care, and getting to the heart of patient and family suffering. At the point when modern medical care can no longer offer reasonable options for treatment, we must rely on traditional person-to-person communication.

In working with people during what may be the worst possible times of their lives, compassion cannot be an added extra—it must be an absolute. No matter the skill you choose to utilize from this text or the myriad talents you may merge and make your own, you must keep your vow of values, keep your vocation, and let your patients keep their story. What does all this mean? It means that it becomes more important to humanize your role as a provider while treating patients at the end of life more than any other stage of care. Physicians and the care team are raised with their individual and specific sets of values and experiences. This is part of the ethical role intertwined with the most rudimentary communication skill. It's good to state who you are and where you've been, in the rawest sense of providing. For example, if you hail from lots of experience with death and grief, if you suffer from your own personal illness, or if you have never known loss, that is ok to expose. Following her father's passing, journalist and researcher Ann Neumann set out to learn about a good death. She noted the hush of the conversation that accompanies demise, even within the hospice she volunteered with. Her book beautifully illustrates the challenges in the "sad work" (2016, 11) of the professionals whose environment always encompasses a patient's end and the journey all mortals make to have a good death, only to be hindered by healthcare complexities and shushed communication.

Unlike other times in a treatment phase, when your role as an authoritarian and expert is crucial, this is the time when expertise must be met with exposed vulnerability so the patient knows you as a human as well. What is nonnegotiable is the certainty of the provider and team to know how to talk—and the overwhelming evidence that this ability does not exist.

Dr. Angelo Volandes, author of *The Conversation*, beautifully introduces the deficiency in skill described in this text (Volandes 2015). His unique book should line the bookshelves of every outpatient facility in the treatment industry, as it eloquently gives guidance to those patients and families who are very unfortunately forced to prompt a conversation that physicians and providers have fallen short at initiating.

> Talking to patients is given short shrift in medical training. The focus of medical education is on technology and treatments; medicine is about doing, not talking. Communicating with patients, especially about end-of-life care, usually takes a backseat....
>
> When I completed residency, in order to become a board-certified physician I was required to prove my competence with inserting central line catheters, leading Code Blues, performing lumbar punctures, drawing blood, and obtaining arterial blood gas samples. But not a single senior physician needed to certify that I could actually speak to patients about medical care. Ironically, I have not inserted a central line or performed many of the other tested procedures since residency, but I speak to patients and families daily. (2015, 27)
>
> In 2008, a group of Dartmouth researchers surveyed all 128 U.S. medical schools regarding offerings in "Palliative and Hospice Care." Of the forty-eight medical schools that responded to the survey, only fourteen had a required course and only nine had a mandatory rotation for students that were interested in the topic. The researchers concluded that only a small fraction of U.S medical schools required training in communicating with patients with advanced, incurable conditions. (2015, 29)
>
> Doctors are by nature defensive when it comes to talking about death, which is of little surprise in a profession where death's presence is the elephant in the room. (2015, 105)

I created a similar study, expanding on the one cited by Dr. Volandes, and surveyed 457 medical school and allied health professional programs around the world, asking a single question: "Do you have a text on end-of-life communication in your curriculum?" With a 20% response rate, 90% of those who responded answered they did not address this area with any real focus and did not have a text to support the skill. One of the programs surveyed replied that teaching nurses to take hope away from a dying patient would not align with their role as advocates. That feedback begs the question: what sense of false reality are we teaching new clinicians, if facing death aligns with removing all hope? There is hope even in death: hope for closure, hope for mended relationships, hope for last bucket-list items fulfilled, hope for peace, hope for a full spiritual cup, hope for little pain and suffering, hope for a left legacy and a beautifully planned memorial to celebrate life. In fact, patients who embrace this hope cite the best times in their life at the end (Berman 2015).

My study concluded that few professional programs in healthcare outside of the specialized palliative and hospice fields are teaching how to communicate in these situations, despite an eventual 100% mortality rate for every patient across the globe.

5.2   Developing Communication Skills

In the remainder of this text, varying skills are described in detailed format, complete with scripting and specific action for you to emulate, consider, process, and most importantly, utilize in some variation in your practice. Some skills are defined with case examples, other with tools for action. Specific actions and instructions are given with brief case examples to follow so that the action is clear.

If at the end of your training, you have not yet conquered the skill and truly cringe at the thought of end-of-life conversations, the very admittance of your deficit is admirable and the only unethical action would be to abandon the further development of that skill. This is when strategy would call for the following: Keep *The Conversation* on your shelves, always be willing without hesitation to bring Palliative Care personnel in early, at the point of diagnoses on serious illnesses or soon after, and focus on that weakness by employing someone in your office—a nurse or extender, a peer or social worker—who is well equipped with the emotional skill set to handle these situations successfully and with compassion. Be patient with yourself.

5.2.1   Basic Communication

VitalTalk (www.vitaltalk.org) is a starting place for any clinician who feels he or she lacks the basic communication skills to serve as a foundation for practicing and refining end-of-life communication (Back et al.).

VitalTalk offers "quick guides"—one-page scripts and protocols to help serve as a framework for skillful communication. These scripts lack the personal touch that real, effective communication demands, but they serve as a place to start. VitalTalk also offers recorded examples of conversations between clinician and patient, both audio and video. These resources can be used to address basic communication skills necessary for effective communication to be cultivated through practice and implementation of the skills detailed in this chapter.

Systematic changes and changes in clinician education are necessary to facilitate effective communication between clinicians and their patients. An emphasis on mindfulness, self-care, and primary care's responsibility for initiating patient communication can help foster an environment conducive to effective communication. (See, for example, Beckman et al. 2012 "The Impact of a Program of Mindful Communication on Primary Care Physicians.") Patients rely on their clinicians to tell them when the end is near. Unless they hear it from the doctor or other trusted clinician, they may incorrectly assume they have plenty of time, and physicians wait too long (Lowery et al. 2013 "Living with Advanced Heart Failure or COPD"). Based on a review of the literature, primary care clinicians play an important and often neglected role in effective communication at end of life or in the event of serious illness. Barriers in the healthcare system may also prevent clinicians from effectively communicating with patients. Improved training and tools are necessary to encourage primary care clinicians to take a larger role in communication and help clinicians navigate system failures that keep them from clear communication (see Lakin et al. 2016 "Improving Communication About Serious Illness In Primary Care").

5.2.2 Skill 1: Professional to Professional

The first crucial skill set is easily lost when ego or intimidation compromises basic communication. This is the language, or lack of it, that occurs between providers or between the care team with the provider. It is crucial that we all talk with each other. Many times, after years of seeing these dilemmas, I am shocked to find two consulting physicians have only relied on notes in the medical record to communicate with each other, and sometimes not even reading those notes in their entirety. We all have cell phones. We tend to guard those numbers as healthcare providers with utmost and at times ridiculously protected discretion, though I have always given mine to every patient and family; only once out of hundreds has a family abused that privilege. However, even when the number is not given to the patients, you should still be reachable by your fellow providers.

When notes are signed in the record, include your contact number for "questions and discussion." Two cases convinced me of this need. One involved a gastroenterology (GI) specialist seeing an HIV patient for a GI bleed. I was called for end-of-life discussion and "futility" on this patient. The infectious disease (ID) specialist who saw this patient regularly and had better insight into the disease had seen the patient only at the beginning of admission. The GI provider did not contact the ID physician to discuss the patient, and thus end of life was pursued. I felt it crucial to explain to the ID physician that I was consulting and was concerned about the deterioration of this patient. The ID specialist disagreed that things were futile, came back in knowing that this patient had stopped a medication to avoid specific side effects, and quickly was able to treat and heal the patient. When I reported this to the GI provider, he stated, "Oh, well, if he says she will recover, that is fine."

Why didn't the GI physician get in touch with the ID specialist directly when he realized things weren't going well for the patient? What is the fear? Are physicians so intimidated by one another, so overworked and without time for a 5 min interaction, or simply afraid of another's opinion? The time argument must stand moot because I spend ten-plus hours on cases that could have been resolved by a 5 min interaction. It is worth the time! I cannot provide any evidence-based certainty. I can only advocate for the skill of the call and the approachability left in the simple sentence "Please call my cell phone to discuss this patient further or for any questions," and provide that number! It is that easy.

The more obvious disconnect in professional-to-professional communication is between the provider and the support staff. Frequently, allied professionals (nurses, respiratory therapists, physical therapists, nutritionists, speech therapists, social workers, and so on) avoid contact and direct communication with the physician/ provider simply because of inherited and learned hierarchal standards and/or authoritarian intimidation. I realize allied teams are taught that the meat of diagnosis, prognosis, and plan of care should be sought and carried out by the physician alone. However, in an age where more extenders tend to see patients for days at a time and where physicians are limited by skill and time, it has become almost unsafe to trust that this communication will occur in our expected way. When a piece is missed, when a conversation is not occurring or patient/surrogate questions are left unanswered, I encourage two tactics. The first is the simple personal pat on the back that you—every one of you—are an expert too. You are the nurse, who sits at bedside 12 h and learns every detail about the patient. You are the social worker who knows more about discharge planning and legal documents than other staff. You are the respiratory therapist who likely knows better how to intubate/extubate/take off paralytics and what to expect each time. You are the speech therapist who makes recommendations to allow/disallow eating; if the ninety-two-year old demented patient will likely pull the tube, *convey your knowledge*. You are the physical therapist who knows if a rehabilitation order is written in vain because the patient will not rehabilitate physically and to do so is to suffer. You are the nutritionist who knows diet better than *any* other healthcare worker. Trust your judgment. Providers may have the MD or advanced degree, but they may not have *your* knowledge or *your* experience.

These trifling communications leave too much space for medical error and lost time at end of life. Write candidly and without discretion in your notes, and go beyond your notes. It is likely that some of your notes are never read. Thus, when you have a concern, reach out, speak to the provider, and ask. If they refuse to listen or to improve the communication, but you notice they are writing more candidly than they bother to relay to the patient, *read their note to the patient/surrogate*. If a stage 3 cancer patient with complications is told, "More therapy can begin when you improve," and the note states, "This patient is unlikely to benefit from curative care and is informed the treatment is palliative; prognosis poor," read that note to the patient. You have not stepped outside your role. You have attempted to ask for clarification from the provider and then read aloud what is written.

Lastly, as an allied team member or physician, do not disregard unique patient aspects in your report. It is easy to become vested in the way a report is passed on historically, the diagnoses, the medications, and the basic job-related requirements from one shift round to the next. However, those unique factors—such as family dynamics and whether or not POA paperwork is on file—can make or break down the communication.

5.2.2.1 Case Example

- My patient was a sixty-three-year-old Caucasian female with stage 4 laryngeal carcinoma, per history. She was found collapsed at home by a friend and brought to the ED. At the hospital, she was assessed to be severely depressed, dehydrated, malnourished, and purposely not speaking to medical staff. She was living with PEG placement and, though independent, had made suicidal choices at home. The MD read "stage 4" and made no call to the oncologist; he recommended hospice. Social workers felt psych should be brought in but never relayed that to the MD. Ethics was called when the patient's sister disagreed with bringing in hospice, stating, "I thought she was almost recovered and just having a difficult time adjusting to family issues." Speech therapy made note that it seemed the artificial nutrition (commonly called PEG) was unneeded but did not directly communicate that to anyone, just documented it in the notes. Ethics recommended psych and called the MD about a PEG tube removal recommendation.

- A psych evaluation revealed that the patient had attempted suicide by putting antifreeze in her PEG tube. The patient reported an onset of depression 2 weeks prior, following the death of her partner of 16 years.
- Ethics spoke with the patient, who stated she was "ready to die." Ethics spoke with Oncology, who stated the patient was in full remission, cancer free, and on the mend requiring no further treatment. PEG was only to increase nutrition. Patient was relatively well.
- Based on physician recommendation and psych evaluation, the patient was sent for involuntary inpatient psychiatric treatment (1013 form involved), and further counseling allowed the patient to resume normal life.

What can you, as an allied professional do? Here are some specific instructions:

1. Start putting a short note on your charts as outlined above. If you are hesitant to use your cell phone number, give your office number—and make sure you check in often.
2. Spend time with an end-of-life patient who is under your care; ask what you, as a stranger, can do for them. Sometimes small things can make a big difference.
3. Make sure you have communicated clearly your own EOL plans to your loved ones.
4. Do not be intimidated to reach out to any professional and ask a question. A 5 min conversation can replace days of angst and will build rapport.
5. From specialist to specialist, remember there may be differing opinions on how close to end of life a patient is and when to draw boundaries. It is best to discuss together before conversations with the family. If you simply offer your counteropinion, you leave distress, guilt, and uncertainty for the patient and family. There may be a middle ground you can find with your colleague before this angst is fueled.

5.2.3 Skill 2: Starting the Conversation Early

Why is it only moderately comfortable to talk end of life when death is obvious and imminent? When a patient might live a little longer with a trach, might make it to nursing home admission, or might get off the vent, we do not routinely consider the end of life. However, even if they do improve, this might still be near the end of their life.

A change of perspective would help as a first step. End of life should not be scary just because it is thought of or defined as imminent. Hopefully, it is a phase, not just one scene. Daniel's end of life was ongoing and lasted years. Technology and advancements allow this, though even without artificial support, end of life can still last an extended period of time. Why is this important? Because it gives the providers and supportive team the permission to start talking about it early.

I can recall a nurse contacting me as her patient was dwindling in the ICU. The patient had advanced stroke, dementia, cancer, and a lengthy list of other more minor comorbidities. The patient was on the vent at about 80%. I remember speaking with the family and utilizing the words "dying" and "end of life." The family rushed back to the unit distraught. It was the nurse who was most confused. She asked me, "Did you tell them the patient was end of life, because he is stable?" I explained that I did in fact say end of life because the patient was end of life. I hoped she had not undone the work I had begun by informing otherwise. This patient died weeks later in hospice. He was not imminently dying, but it was important for the family to accept the phase. For most, the acceptance does occur for hospice services in last days or weeks; however, this is partially dependent on race: African Americans favor hospice much less of the time (Varney 2015; Benton et al. 2015).

Physicians have to be able to recognize when diagnoses are adding up to equal a patient's last phase. Only the will and spirit of the patient will define when imminence is present, but the hospital is usually when it is just too late. As a rule of thumb, if the patient is admitted and on artificial support, the conversation is a must.

However, even when the patient is under your care, outpatient, and diagnosed with a serious, eventually terminal illness, the conversation is at a good starting point. The hospital tends to be too scary anyway, so to begin when the patient is relatively well is a best scenario.

5.2.3.1 Case Example

- The patient was an sixty-eight-year-old Caucasian male who presented to the ED with shortness of breath. A month prior, he had been hospitalized for treatment of a urinary tract infection with antibiotics. In the ER, he was intubated and admitted to the ICU after a CT scan from his previous hospitalization showed an irregular nodule on his lower left lung.
- At the time of ICU admission, the patient's niece was his POA and he was had filed a DNR. However, his niece stated she wanted everything done.
- Ethics was called to speak with the niece, who was seemingly making decisions against the patient's wishes.
- Since the patient had been to the hospital a month earlier, a living will could have been filled out during that visit, so a conversation with the POA might clear up confusion and ensure decision making in line with the patient's wishes.
- Ethics discussed the patient's wishes with the niece/POA and explained the need for DNR/comfort care if that was in line with what her uncle would want. Ethics requested that the POA check to see if the patient had a living will to give clarity to his wishes.
- Niece decided DNR was in line with her uncle's wishes and changed code status.

5.2.3.2 Case Example

- The patient was a sixty-two-year-old African American female who returned to the ED shortly after a recent discharge. During her prior hospitalization, she had undergone a tracheostomy and was sent home on a 40% aerosol tracheostomy collar and tube feedings. Her husband brought her back to ED after concern that he could not get her O_2 saturation above 40%. She was admitted to the ICU and placed on the vent.
- Ethics was reconsulted on the case after the husband refused a hospice consult. The husband was insisting that he would care for the patient at home. An RN stated that the husband had unrealistic goals of care and, in the meantime, the patient was suffering.
- Ethics was on the case for 2 weeks and had multiple discussions with the husband, who was undecided. He finally agreed to fulfill his wife's wishes to be at home by having a hospice agency transfer her home and extubate her once she arrived at home.
- If an EOL discussion had taken place earlier, the husband would have had a better understanding of his wife's wishes in the face of suffering, and he could have made a faster decision, saving the patient from 2 weeks of suffering without comfort care.

5.2.3.3 Case Example

- The patient was a forty-three-year-old Caucasian female with advanced, neglected, systemically metastatic adenocarcinoma of the breast. She presented to the ED with diabetic ketoacidosis and infected ulcerating breast legions.
- During a frank discussion with an MD about the patient's condition, her husband reported that his wife was fearful of having a mastectomy and had refused to return for any doctor visits for the past 6 months. Due to the severity of the neglect (one breast was assessed to be completely replaced with cancerous tissue and extensive metastatic carcinoma to the spine), hospice and palliative consults were the most appropriate option.
- Ethics spoke with the husband to address hospice/palliative options. The husband revealed communication gaps between he and his wife, whose biggest concerns with hospice were logistical. Few clinicians understand a dying patient requires 24/7 care and oftentimes family members cannot or are afraid to help with this at home.
- If the patient and husband had spoken when the patient had received a diagnosis, a realistic plan of care could have been established to respect her wishes and fears. Instead, the patient and her husband neglected diagnosis and EOL conversations, leaving the patient with no options.

What does it mean to start a conversation earlier than imminent death or even hospital admission? Use the following tactics:

1. Make certain the entire family understands and is educated on the progression of the illness. For example, in a patient with COPD, it would be best to plan an education session, ask that important family members attend, and explain that this disease has a beginning and an end. Explain that the care team will treat and prolong for as long their loved one's quality of life is realistic, but that they need to understand what is unrealistic for the patient.
2. Recommend at this education session that a nurse or staff member assist with advance directive completion so that those witnessing the patient set his or her own quality-of-life boundaries can also be the ones to carry out those decisions.
3. As the disease progresses, consider allowing a palliative specialist to follow on an outpatient basis.
4. On each appointment, continue to use language that embraces the journey of this disease, such as, "You seem to be doing well, but the disease is progressing as expected. Let's keep living well with medication and treatment, and remember that the time will come when boundary setting for those machines is needed." This way, when the terminal state sets in, there is not such a shock.
5. Remember that these conversations require repetition, compassion, and continued hope to allow processing to take place. The patient should always leave with a positive goal in his or her life but should not leave with a false sense of what the disease is.

5.2.4   Skill 3: The Nut Graph of the Conversation

As professionals, you need to understand the tactics, actions, and strategies that will lead to a successful outcome of an end-of-life conversation. The first step in doing this is to define what your successes or your goals should be. If you have unrealistic expectations, you may feel like your attempt to step out of your comfort zone and breech the barrier from treatment to something less than full recovery is a failure. It is never a failure if it is communicated. However, it sometimes must be said many, many times before people can process that they or their loved ones are nearing life's end.

After I have a conversation with their patient, many providers ask, "How did that go so differently? I said just what you said." That very pertinent question represents a paradox and an answer to a good inquiry. How does that happen? Many theories come to mind. Sometimes it was that eighth time that made the difference. Sometimes it was the eye-level communication. Sometimes it was the compassion. Both repetition and ability to process may change the same words to a difference in understanding. The process that I outline here is intricate, almost procedural, and should be memorized and followed for better outcomes. The obvious tactics are timing and motivation to listen.

1. *Timing.* If your attempt follows several failed attempts to half-heartedly declare the end is near, you will be drowned out by those who came before you. Best to wait and begin rapport, define yourself as someone different at that point. This might mean waiting a day, just to give the patient a break from the end talk. If an event has occurred with another hospital or opinion that has created an air of distrust in all healthcare, give the patient time to process what has been explained. This internal processing time for the family may increase the length of time it takes for them to come to trust you. They must address and resolve the issues causing their earlier distrust before any discussion with you is worthwhile. The conversation is lost on emergent situations as well, since the family will be frantic and the patient is usually uninvolved at this point.

   However, in a family who is unwilling to change to a DNR, witnessing their loved one's code may be a good time to readdress this situation. The answer is in the process, not the pressure. Likewise, you may be there. You may see the futility and wish the family could see it as well. What makes this process so difficult is the journey they were on before they met you. Unless very progressive doctors took them by the hand early, they have likely heard next steps, next treatments, how things will be fixed, and what aggressive escalation is recommended. Then all of sudden, from their point of view, there is no recommendation but the withdrawal of everything they signed consent to. That understanding does not happen easily. And likewise, leaving too much time to follow-through with an end-of-life plan is not a good tactic. Time kills decisions.

2. *Sit down.* It's been written before and it is no joke. Eye level or below is key. Frequently when I cannot locate a chair in a crowded hospital room, I sit on the bed with the patient, or I squat down. I do not want any intimidation to be conveyed through my body language. They need to think clearly so questions can be asked and answered. All too often in healthcare the patient is spoken over, not with.

3. *Listen.* The healthcare rumor mill is sometimes frivolous for no reason, but that does not excuse the need to get to know the dynamics you are stepping into it. You might use those details and dynamics to choose the strategy of explanation. It might also build instant report with the family to know something of their psychosocial situation. For example, "I understand you have been caring for mom for eleven years; you are a good daughter." Be motivated to listen. It will *save* time to *take* time. Again, it will *save* time to *take* time. It is unnecessary to walk in the room while looking at the chart; leave that to the doctors on television. Such an action immediately disconnects the patient/surrogate from your expertise. If you do not have the whole picture, be sure they can give it to you. Some of the best conversations begin: "Hi, Ms. Smith, I am Dr. Benton, Kathleen Benton. Before I start talking I would like to understand what you are thinking. Tell me what you understand about your illness." Do not make too many notes, other than those disconnects you need to clear up *after* the patient or family finishes talking. This very simple communication may eliminate weeks of frustrated discussion.

4. *Simplicity is in the semantics.* Using your patient's and families' names and attempting to hold on to that information will help your communication. The jig is already up about the very complex and overly burdensome clinical language hindering lay communication, but there continues to exist great confusion on how challenging what we consider basic terminology seems to be for those who have never been in medicine. Change DNR ("Do Not Resuscitate") to "Allow natural death," "Life-support" language should replace "vent" or "respirator." All organs should be referenced as they are understood by the patient: not pulmonary, but lung; not renal, but kidney; not cardiac, but heart. When only one person understands what is being said, it is not a conversation but a lecture. Some of the most challenging semantics that can break a conversation include the term "brain death." If the patient isn't dead, don't say it, or you leave space for outsiders to argue "reflexive activity." Likewise, don't say "nothing we can do" or "not much else to add." There is always something to add; it just may be comfort and hope for legacy.

5. *Do not speak only to your organ.* You are a well-trained professional. Despite your expertise in that one area, you should know that having good kidneys in a metastatic patient does not equal "things are better." Many patients and their families are confused by the positive information given to the recovery of one organ, when many of the others are dwindling. It is misleading and inaccurate. Give a big picture, something like: "She is very weak today. Even with her kidney numbers recovering, we know the cancer continues to weaken her, and she cannot live on kidneys alone. We need to look at the whole body."

6. What to say is just as important as what not to say. If you have difficulty, practice saying some of these aloud:
   (a) Use the phrase "at end of life," followed by quickly explaining this could mean days or months.
   (b) Use the word "dying," because if the patient is dying, it needs to be said.
   (c) Ask the patient, "What can I do to support you?" The patient and/or family is experiencing anticipatory grief; be their ally and recognize their devastation. I have personally offered to bring food to a hospice or home, even when they leave my institution. You aren't bound by the four walls, so reach out, order flowers or food if that's what you feel called to do.

5.2.5   Skill 4: Compassion Is Learned

Not everyone is born with compassion and empathy, and it is a difficult value to teach or learn. But as with anything, you can fake it until you make it. Offering support through words outside of what it means to be clinical can be a sign of that compassion. Even if it feels uncomfortable at first, you will find it becomes natural and innate and you develop the ability to see from a perspective of someone who is hurting so deeply you can only feel for them.

1. *Remove personal guilt.* Neither you nor your patients and family are guilty or responsible for death, particularly in your chronically ill patients who have been

cared for by family for many years. Paradoxically, in those families who have not cared for a family member and instead moved away while loved ones got sicker, personal guilt will rain thick. Tell them that illness defined the moment, not caretaking or living their lives. Now is not the time to allow guilt to prompt trying desperately to preserve and prolong their loved one's life but rather to selflessly let them go because they cannot be saved. "You are not choosing if your loved one lives or dies. That is not why we are here. You are choosing how they would like to spend their end of life. You are not the catalyst of death, the disease progression is. Despite what is done, barring a miracle, they are dying. Do they want to die on artificial support or naturally?"

2. *Let the patient/family salvage some control.* In dying, we lose all our control, and a family is useless to a dying patient in the hospital—or that is at least how they feel. If the family or patient needs more transparency or to see a record, if they need to vent, or change the terrible food delivered, let them. These are the battles better surrendered to win the war on peace and dignity.

3. *The element of suffering cannot be ignored.* Is pain relevant? Does talking about it enhance the guilt a loved one may feel for choosing aggressive care? Yes to both. Suffering is not just pain. It is burden of care, constant complications, dignity lost when the inability to communicate is absolute. Talk about suffering and the possible resolve through palliation. Suffering should not be an abandoned part of the conversation, nor should it be the catalyst to euthanasia and assisted suicide when true communication may inhibit this possible outcome. Some patients are afraid to say they hurt. Some families do not know they hurt. Many believe the act of withdrawing aggressive care may cause suffocation and suffering. Communicate realistic expectations.

4. *Things are not black and white in the realm of end of life.* They are gray. What was a terminal cause of death many years ago might now be survivable. You can choose to minimize support and not completely remove it; if you do remove all support, it does not necessarily mean that life is over. Longevity and quality of life without suffering through palliative measures is an option. Referencing the Chap. 3 distinction between palliative and hospice care, a provider can allow a personal, customized plan of care before death. For example, a patient who wants longevity with a trach and vent can choose to stay on them but may feel burdened by dialysis and decide to let go of that technology.

5. *Walk with your patients through the death, or bring in a hospice that will.* Once withdrawal from all measures is decided, the hard part is making sure a family still feels they have made the right decision when survival continues. No one can know for sure when a patient will pass. I recommend bringing in hospice to cover this base, because death is their organ. If you do not, loved ones might start to wonder, for example, whether they should force feed or allow IV fluids. Consider their angst and make sure it is all addressed. Again, this is that time to offer more support, even if they do not wish your extras. "Can I visit you at hospice?" "Does your family need our patient advocate to help with funeral planning?" Many are lost during this time. You do not have to be their expert, but you can find them some guidance.

5.2.6 Skill 5: Owning the Discussion

There is not always a next, escalating step in a treatment plan. That should be clear by now. But there is always a reason to stay in the game as a provider. Your "next step" may feel very unlike all you were trained to do. Let's review paternalism, the positive act of guidance from a physician expert; paternalism is using intuition and discernment to decide which of the many available treatments and prescriptions to include for this particular patient. It is medicine as an art form. Paternalism at its very definition is positive—think "parent-guider" or "adviser." It has gotten a bad rap in recent years as we edge toward more autonomy, yet this is where balance is best.

When do all the facts, charts, and clinical protocols (the science) become enough and the time for discernment (the art) become apparent? Does judgment even exist any longer or do we practice using only a checklist? Certainly a physician uses his professional acumen when choosing the right chemo or antibiotic, and this same ability must apply to the end of life. You need the ability to offer things and to say no. Instead of roundtables to discuss the "why" for all this chosen futility in health-care, consider the population of invincible ignorance we are creating through our intelligent denial of the obvious problem at hand. No matter what technology we lend the body, it is only a rental until our mortality is realized. There are those family members who will never escape the ignorance of presumed immortality our profession has created. Invincible ignorance applies to those who truly feel death is something we can always avoid or ward off, and who refuse to admit otherwise.

1. All options do not have to be presented. A surgery is not always the best option just because it is available and your mortality rate will stay alive because you can get them through it. If an elderly nursing home patient loses weight after they stop eating, a PEG does not have to be the solution. If life after that surgery is harder and re-admission is likely, should it be an option? Will you use your judgment in end of life the same as you use it when treating an evident diagnosis? Will you be able to embrace the belief that less is more? Will it make sense to take a BiPap from a dying patient and replace with a palliative order for morphine, knowing the morphine sustains quality of life while the BiPap offers length of life? If you don't, and you choose to be less paternalistic, the family and patient will suffer. To preserve your own integrity, the patient and family must feel they fully own the decision. So how do you admit to and recognize death when it feels more intuitive than diagnostic? We are so wrapped up in those checklists that our gut and intuition does not play a part.
2. Offer a hybrid approach. A physician friend suggests an example of this: "Ask the patient, 'What is your definition of successful treatment?' I tell them that if the doctor tells you they are going to put you through hell and cure you, that is very different from they are going to put you through hell for the possibility of a little more time alive. It is not easy to tell patients the truth, as you well know, but I have never regretted it."

3. Expose your humanity. How do you practice paternalism in treating the end-of-life patients? You expose your human side and dig in to your intuition. Being raw and vulnerable is the most difficult part of being human, especially as a professional. Exercising one's right to be raw as a professional seems to go against the very essence of professionalism. I prescribe to a different line of thinking in this area of medicine. Your patient's end-of-life phase is the time to expose your feelings, be vulnerable, and practice being as human as you can. This may be the time to tell the patient where you come from, what your beliefs are, and risk those internal boundaries to get close. Normalize this by sharing with peers your own experience. If we disassociate, we disconnect. Did free will or compulsion bring you to your profession? If by free will, there was probably a driving experience in your life that prompted it. Be that person in these scenarios.

For me, the share is obvious. When I feel a family needs to relate and believe, I empathize, I share my Daniel. I have had countless patients and families thank me, hug me, and relax when I share. He is my story, my inspiration, and what makes me human, what fuels my ability to own this discussion, where I know more than they about the outcome and can lend some insight to remove their burden. This is not a tool that can be scripted; tell what you know, tell what you have seen. Do not fear you are intimidating or pressuring by speaking to pain at the end when medical advancements are overused.

5.2.7 Skill 6: Recognizing Cultural Barriers

Cultural competence is not simply a buzzword created in the healthcare realm. It is a valid concept that must be understood and respected. Individuals from some cultures need to die thinking they will never "die." Some cultures need to protect the dying from information, and the patient seeks this protection. Some will wait for the miracle through the very end state. Others need to seek counsel from elders and outside authority. Invite those in. We all process information differently, and you cannot know how it feels to have that skin color, background, or socioeconomic status, to be judged or to feel distrust. You cannot necessarily empathize, but you can meet them where they are. Some only process a piece at a time, some in bulk with the support of others. Your professional role is to educate yourself on *their* choice by asking questions.

5.2.7.1 Case Example

- The patient was fifty-eight-year-old Hispanic female who suffered a stroke. Her husband deferred communication and decision making to his seventeen-year-old daughter. The RN originally assumed the husband did not speak English but later found that he could understand a small amount.
- Staff was speaking with the daughter directly, who was scared and emotional and could not comfortably relay the poor prognosis to her father.

- Ethics did not support the deferral to a minor and made sure a translation device was being used by the staff and MD to aid in communication without the help of a minor.
- Ethics also encouraged the use of a translator to facilitate good communication between the husband and the staff/MD. Once the husband understood the devastating neurological situation, he was able to relay to the clinicians that the patient would not choose this quality of life and asked that they withdraw life support.

5.2.7.2 Case Example

- The patient was a seventy-eight-year-old Hispanic female admitted for abdominal discomfort caused by necrotizing pancreatitis.
- Ethics was called by the MD about concerns that "cultural barriers" would inhibit family decision making and informed recommended care plan for the patient.
- Ethics spoke with the family, who stated they all understood and could speak English. They all understood the current prognosis and knew the patient's wishes for a natural death and peace.
- The staff/MD assumed a cultural barrier with this family when one did not exist. The family was easily able to understand the prognosis, relay the patient's wishes, and make appropriate code status/plan-of-care decisions.

5.2.7.3 Case Example

- This forty-two-year-old Indian patient was admitted to the ICU with a gastrointestinal bleed. He had a previous history of cirrhosis, presumably from being an alcoholic. He presented at the ED with decreased mental status and vomiting. He was intubated upon arrival in the ED and laboratory work revealed significant metabolic acidosis, severe anemia, and hypotension. The patient had supposedly quit drinking 2 years prior to his admission.
- The patient had been hospitalized five times in the past 6 months for issues of volume overload and anemia. He had received multiple blood transfusions. Patient was also uninsured and listed as a self-payer.
- Ethics was called due to the physician's concern that the patient's care was futile. Also, the patient's wife was not making decisions until she could confer with other family members back in India. She would not allow the MD to speak with the family, and there was concern that she was relaying the medical information about her husband's prognosis incorrectly.
- Ethics spoke with the very frustrated wife, who explained that the patient's brothers in India must be the primary decision makers. Culturally, the wife was following her family culture of allowing the males to make decisions. She was also trying to withhold the information about her husband's alcoholism from the family. She was unwilling to let Ethics or the MD speak with family in India for fear of divulging the patient's history of alcohol abuse.
- Per the MD, the wife had received advice from a family friend who also happened to be a physician that she should continue with aggressive treatment.

During the Ethics conversation, the wife was unwilling to change the patient's code status. However, the wife did seem to understand that her husband's care was futile and the recipient of burdensome care.

- Ethics advised the MD to continue communication with the wife and seeking the patient's wishes, but the MD should not pursue any treatment that made him concerned or uncomfortable.

Some of the tactics you might try include the following:

1. Learn a few things about the culture; simply ask if you do not want to do the research
2. Try to be objective to their way of life. It is their values, and recognizing and empathizing with them will help you care.
3. Always bring in a translator no matter the time it takes. Never utilize family to communicate. They are too emotional.
4. Ask yourself if you have your own judgments of their culture and try to avoid bringing them into your conversation. Think beyond the typical race, gender, and so forth. Do you automatically assume obese people make bad food choices? Do you assume non-English speakers have no American healthcare education? Work on challenging your assumptions. Get beyond your own cultural comfort zones. Be open to difference.

5.2.8 Skill 7: Defining Privacy Versus the Need to Be Informed

Many patients wish to keep their illness private and guard their diagnoses, especially when it is something as sensitive as a terminal issue or one more stigmatized, such as HIV.

5.2.8.1 Case Example

- The patient was a thirty-five-year-old African American male admitted to the hospital for heart failure.
- Ethics was called because the family was not aware of the patient's HIV diagnosis. The MD was concerned about whether to inform the family in order for them to make decisions.
- The patient's mother was his decision maker/caretaker and was aware of his HIV status, but the patient had teenage children who were not aware.
- Ethics decided to protect the patient's privacy since the mother was a fully informed decision maker. The patient's children did not need to know about his HIV status since they deferred to the patient's mother and were not making decisions.
- The patient's children were informed about the severity of their father's condition without including information about his HIV. The family supported the mother's decision to withdraw and allow a natural death.

5.2.8.2 Case Example

- The patient was a fifty-nine-year-old Caucasian male who presented to the Emergency Department with leg weakness and cramping. He had a history of cancer but had decided against treatment, wishing to carry out his end of life in his own natural way.
- Ethics was called by MD with issues concerning diagnosis and questions about informing his family. Because the patient had kept the diagnosis to himself in the past, Ethics supported maintaining his confidentially until an issue with inappropriate or nonbeneficial care arose.
- As the patient prognosis digressed to "poor," Ethics supported the need to inform the family in order to make decisions for the patient. The family had begun violating the patient's documented living will wishes to allow natural death and had rescinded his DNR.
- At a family meeting, Ethics spoke with patient's parents and son only. Other family members showed up, but those individuals were not included in the meeting to maintain privacy.
- Ethics explained the patient's choice against disclosure and treatment by bringing in documented records from oncology. All were shocked, but all three agreed that the patient would want DNR/WD in order to have a natural death.
- The rest of the family members were informed of the severity of the status and the family's wish for DNR or withdrawal orders. The patient was an active minister, and the family requested confidentially.

As a guideline for you when dealing with family members, a difficult conversation about what to do next requires two sets of formidable questions answered first.

(a) Is the patient able to speak for himself/herself? Can he or she make decisions? Can he or she be advised to choose a POA as confidant to make those decisions when the patient is no longer able to do so?
(b) If patient is not alert, is the terminal information necessary for the surrogate to make a truly informed decision?

In any discussion with family members when their loved one has requested privacy, reveal only the information necessary to make a decision.

5.2.9   Skill 8: Steering Clear of False Hope

Let us dive into the concept of false hope. As a professional, it is not acceptable for you to create false hope during the discharge of your duties. This is to be distinguished from the type of hope that denial allows, which is acceptable as a coping mechanism. The false hope should only come from within the patient or family and be allowed after the attempt to inform has occurred and patient has chosen the emotional path of least resistance. Some people need to die believing they will never pass, and some need to hold onto hope for whatever that means. If you offer the facts and do not ask them to sustain or choose an alternative that will provide the same outcome, you are being fair to their spirit. If you only give the ethical options (sometimes when care is futile, this means no options but comfort), then you have done your job.

5.2.9.1 Case Example

- The patient was a seventy-one-year-old African American male. He presented to the ED complaining of abdominal pain. He had a known diagnosis of metastatic bladder cancer with lymphatic involvement and was currently undergoing chemotherapy at another hospital.
- The patient and his wife told the ED physician that the patient had been doing well since his last hospital admission where he had finished antibiotic treatment for a urinary tract infection. The ED determined that the patient likely had a small bowel obstruction secondary to a mass in the proximal ascending colon, and he was admitted for possible surgery to remove the mass.
- Ethics was called by the MD to address code status after the patient admitted as a full code. Ethics spoke with the patient's wife and son. The son was understanding and receptive to an end-of-life discussion and possible hospice. The wife was very upset when approached about the end of her husband's life, consistently stating, "The oncologist said he was doing well." This was likely the case, but the comment had probably been directed at tolerating chemo, not his overall health.
- The wife claimed that no physician had ever brought up end of life. She understood the value of DNR and natural death, but wanted to hear a doctor she knew and trusted explain that her husband was at that place before she changed his code status.
- After a conversation with the patient's doctor, the wife was willing to sign hospice orders and pursue comfort care for her husband.

5.2.9.2 Case Example

- The patient was a sixty-three-year-old Caucasian male. He was admitted to the hospital for recurring abdominal pain following a below-the-knee amputation a few days prior to this admission.
- Ethics spoke with the patient's wife, who wanted to take the recommendations of the doctors when it came to code status and patient wishes.
- She asked, "Do the doctors all think now is when everything has been done and it is time to move to hospice? I thought they would tell me when, but they never said anything. All their conversations have focused on the positive."
- She simply needed to hear from someone in authority that the patient had reached the end of the road as far as treatment was concerned before deciding on hospice
- The patient was discharged to hospice facility.

Keep in mind the following:

1. Hope can survive if it needs to play a part—but only the hope that is patient created, not physician or care team provided.
2. Look at hope in terms of not something we have but rather something we receive. The mystery of the finite and the infinite are intertwined. From a patient's perspective, hope comes to meet us so we know that we don't just listen to the narrative, we develop the narrative.

5.2.10 Skill 9: Recognizing Capacity

Our mental capacity is task oriented dependent on altered status or mental impairment. A patient may be able to choose a healthcare decision maker but be unable to process what a Do Not Resuscitate would mean for their body, for example. It can be an impediment to end of life, but never an excuse to ignore a patient's autonomy. This is a difficult distinction to make in healthcare. Clinicians and providers, feel so much more comfortable speaking with a person when he or she is off the machines in the room. We would rather discuss the frailty and sensitive nature of end of life with the person whose respirations and vitals are not beeping in our face. Sick people are difficult to converse with, but that does not mean they lack capacity. Ordinary well people are on medications, have bad days, and make bad decisions; that does not mean they lack capacity. And even patients who are confused for moments also have moments where they are lucid and should not be ignored. The best decision maker is the informed patient. If that is at all manageable, it should be sought.

5.2.10.1 Case Example

- The patient was a sixty-three-year-old African American man with an extensive medical history to include cardiac disease with sudden death, from which he was successfully resuscitated.
- He showed up in the ER reporting weakness and abdominal pain and was admitted for stabilization of his hemoglobin.
- Patient left against medical advice.
- Ethics spoke with the patient's brother, who explained that the patient had done the same thing in the past and was acting with complete competency and according to his own free will. He eventually came back for pain related to undiagnosed cancer and was soon septic. The patient again asked to leave.
- Ethics spoke with the patient. He explained that he hated hospitals and only wanted his pain controlled. He did not want medical care. Ethics explained that he would die. He was ok with this reality. He was willing to allow hospice to keep pain under control at home.
- Considering the patient was clearly informed and acting on his own behalf, Ethics had to support his wishes to deny treatment and go home to die.

5.2.10.2 Case Example

- Patient was a fifty-year-old Caucasian male. He had a known history of end-stage renal disease and was on hemodialysis. During his first admission, he presented with an infected tunneled catheter. His MD recommended that the current catheter be removed and a new one be relocated. The patient adamantly refused the treatment, but he allowed the MD to remove the old line and insert a new one in the same location.
- After discharge, the patient returned with complaints of lethargy and feeling that he might fall. After a second admission, it was determined that the new catheter had become infected and the patient had become septic.
- The patient was told he would need dialysis and a new, relocated catheter to treat sepsis. The patient again refused the treatment. He stated the pain was too bad and he was done with continued dialysis.
- Ethics spoke with the MD, who stated that the patient was alert and seemed fully informed when refusing treatment.
- Ethics supported the patient's autonomous wishes.

The following guidelines might help you figure out your most ethical path:

1. The patient should be spoken with, period.
2. If support is needed, look to family and always look for the actual POA document, and not simply the information "So-and-so is the POA." Sometimes, that information is wrong.
3. Even though a patient is sick and vulnerable, unless they state they want to be protected from information, they deserve to be informed.
4. A patient needs to hear "You are dying."

5.2.11 Skill 10: Bringing Together Everyone Who Matters

For some reason, providers and professionals find it somewhat intimidating to enter a room full of overwhelmed and involved family and friends who wish to be a part of an end-of-life discussion. This may correlate with society's overall fear of public speaking or may just be a personal desire to steer clear of drama. Frankly, the presence of many opinions in the room does confuse things and is more work at that moment. However, I always recommend it. If the family is seeking outside advice and they meet with you, and then relay that information (probably forgetting many of the details and remembering others inaccurately) to those they trust and ask advice from, other family members might have heard the details or remembered things more accurately. So do not fear the village. Bring them all, because their opinions will weigh in, in that room or behind other doors where information is convoluted and confused. I always say, "We will need to have a meeting when all those who weigh into care decisions can come and be prepared with questions. Feel free to invite your pastor, your neighbor the nurse, or a cousin from the hospital—whomever is important to the patient's care."

However, I will add, this scenario can be cause for concern when there is a dominant personality who is changing the perception of those in the room. In these scenarios, I directly communicate the issue. I recall a granddaughter who would change the minds of a dying patient's daughters every time they decide to let go. I simply informed her she was adding to the drama and the suffering of the patient, and if she did not listen and support the legal decision makers, she would no longer be included in meetings.

5.2.11.1   Case Example

- The patient was an eighty-two-year-old African American female. The patient became unresponsive at home after having significant nausea and vomiting. The patient was brought to the ED, and a CT revealed a large left-side intracranial hemorrhage with midline shift. She was intubated and placed on a ventilator.
- The patient's medical history consisted of diabetes, two previous strokes, dementia, peripheral neuropathy, hypertension, coronary artery disease, and a remote history of deep vein thrombosis, as well as a possible history of focal motor seizures.
- Ethics was called after the family wanted to purse aggressive treatment (trach/PEG) for a patient who had been determined inappropriate for further aggressive measures.
- Ethics met for a lengthy discussion with the five children. Before the meeting, the children were considering different treatment options, but they were not all on the same page.
- The conversation covered the patient's wishes and burden of care. The family discussed the patient's decision for hospice when another family member had been in a similar medical situation. The family spoke together after the Ethics conversation.
- After the family conversation, one sibling relayed the joint family wishes for DNR and no trach/PEG. They decided to withdraw the patient from life support and place her under hospice for comfort care.

5.2.11.2   Case Example

- The patient was a forty-one-year-old Hispanic female with a history of small-cell lung cancer undergoing chemotherapy. She had a history of chronic respiratory failure, COPD, interstitial lung disease, and lupus.
- After arriving at the hospital, her respiratory status continued to decline and she was placed on a ventilator.
- Ethics met with the husband and multiple family members, which revealed disjointed family dynamics and unrealistic expectations. The patient and husband had been keeping much of the diagnosis and prognosis from the children and other family members, and the husband's goals for the patient were unrealistic.

- During the lengthy conversation, it took multiple redundant statements about coding and life support for the husband to understand that neither would realistically help his wife's situation.
- Other family members were hesitant to join the conversation for fear of upsetting the husband.
- Ethics prompted family members to give their opinion and they told the patient's husband they thought it was time to let go. At the end of the conversation, the family all agreed that DNR and hospice would be appropriate if the patient was able to sustain off life support.

5.2.11.3 Case Example

- The patient was a fifty-year-old Caucasian male. He was in renal failure and ventilator dependent secondary to complications from a sacral decubitus ulcer (for which he had undergone several debridements). His wound failed to heal after months of treatments, and the patient went into respiratory arrest and renal failure during the most recent hospital stay.
- The patient informed staff members that he wanted to be DNR and pursue comfort measures, but he would not relay the information to his children and their spouses, who insisted on continued aggressive treatment.
- Ethics was called to speak with the patient and children after the patient had told the staff that he wanted comfort care only.
- Ethics explained the patient's wishes and encouraged the children to support comfort care decisions. It came up that one child knew his father's wishes; he had relayed them years earlier. Others had many questions.
- After speaking with the children and spouses, Ethics asked that one witness quietly listen to the conversation with the patient. The spouse watched as the patient clearly and coherently informed Ethics that he wished to be DNR/DNI and transfer to a hospice facility.
- The children and spouses were ok with their dad's decision and believed he had been protecting them from knowledge of his illness.

Some of the things to keep in mind with regard to family dynamics:

1. Many (fifteen or even more) in a room is only overwhelming to you, and you are not the focus of the conversation.
2. If you don't inform the whistle blowers or rumor throwers, you run the risk of their controlling the conversation when you are no longer there, undoing your hard work.
3. When there are one or two members who are purposely creating unnecessary drama and they do not have legal decision-making authority, you may need to privately advise them they are not being patient focused and will be excluded from information sessions if they continue to self-focus.

5.2.12 Skill 11: A Note on the Faith-Based Population

Providers cannot exclude God or other Omnipotent Beings from the conversation in a faith-based family. At times, the family can use religion as a crutch and may refuse to make decisions. In relation to patients who die with hope that a cure will come and a miracle will heal them, that is maybe how they need to die, rooted in reality but grounded in some faith that is beyond others' understanding. Hope is qualitative and limits are quantitative. We in medicine are drowning in information and starving for knowledge relative to the underuse of evidence in medicine. Therefore, whether you are a provider who believes some more powerful being or has doubts, your belief in something is important to state before the patient or family even can.

This is the most scripted language I use, because it is always true. I always know when the family is of strong faith because I do my research (simply ask the nurse; they always know). I then begin the conversation, "I do not know your exact spiritual beliefs but I, as a professional, feel it important to disclose to you that everything I am saying today is purely scientific. These are the facts from the tests and experts you trust to care for you/your loved one. We also cannot discount the possibility of a miracle and to recognize we do not have the last say. Therefore, we need to focus more on the burden of care and think about what we may be doing out of God's plan, utilizing our human tools and free will with His plan."

For example, in a Christian family in fear of euthanizing their loved one by removing support, I respond, "I am well versed in the Bible and I know for sure none of Jesus's miracles required machines. We might intervene when Jesus is calling [the patient] home, but if God wants to work a miracle, He will do it with or without our machine. We are not losing faith in God by removing that device. We are instead removing faith of technology and putting all our faith in God."

5.2.12.1 Case Example

- The patient was a sixty-two-year-old Asian male nursing home resident with a history of end-stage liver disease secondary to alcohol cirrhosis. He was brought to the ED with altered mental status and admitted for further evaluation and treatment.
- The patient's wife was told her husband had a very poor prognosis and aggressive treatment would likely not help. The wife refused to consider DNR/comfort measures. She stated that her husband wanted "everything done," and she said that God could perform a miracle if she did not give up on her husband.
- Ethics was called by the MD for appropriate care issues due to the physician's refusal to continue aggressively treating the patient. Ethics spoke to the wife and explained that the patient was in such poor condition, he was no longer a candidate for further treatment. Ethics explained the wife's only option was to fire the MD and find another who would treat her husband. The wife stated she did not wish to fire the MD and she would agree to DNR and hospice if no other options were left; she also advised Ethics she was not "giving up on God and her husband."
- The wife eventually agreed to hospice.

Religious faith is personal, and each family will have its own preferences and traditions. Work with them.

1. Involve a family's religious leader to help them distinguish what is allowable. Most faiths recognize withdrawal as an ok decision.
2. Do not be resigned to excluding miracle and faith; there is some higher power than you, despite your ego or beliefs.
3. The spiritual realm always plays a role, even for those with no religious background; fear of nothing more can be just as powerful.

5.2.13 Skill 12: Learning the Logistics of Discharge

Physicians are responsible for the logistics of discharge. Understanding what you are doing to a patient and how it will affect where they are going next makes a huge impact on what you may decide to offer, particularly if the patient will be discharged with artificial support. We must work harder in healthcare for the left arm to know what the right one is doing. Put simply, we have to know how our knowledge and expertise impacts the next discipline of care after us. If we are not on that same page, the family is often more confused, and the distrust and disconnect in healthcare radiates.

5.2.13.1 Case Example

- The patient was a sixty-year-old Hispanic female with anaplastic large-cell lymphoma. She was admitted to the hospital during a follow-up visit after receiving cancer treatment. At the time of admission, she was experiencing fevers, neutropenia, and significant decline in health over the past week.
- Ethics was called to speak with the patient, who was refusing to discharge to a nursing home although her family was stating they could not care for her at home.
- After trying to accommodate home care, the patient and family eventually agreed on a compromise of short-term nursing home placement.

5.2.13.2 Case Example

- The patient was an seventy-two-year-old African American female in a chronic vegetative state due to massive intracranial hemorrhage. She was on a home vent and receiving home care from her family. She was admitted to the ICU after being brought to the hospital by her family, who suspected she had a urinary tract infection.
- Ethics was called to speak with the family about plan of care for a patient with such a poor quality of life. The patient had been admitted to the hospital about once per month for the past year. Ethics spoke with the family about quality of life versus quantity of life. The family members were not in agreement about

plan of care. Some children wanted to continue aggressive care and vent dependence and others wanted to withdraw from life support.

- For the patient to be discharged and remain on vent, several resources had to work together to include hospice and medical supply. Ethics spoke to several family members to attempt a joint decision but also advised guardianship if some children felt the patient's wishes were not being honored.
- The patient was discharged home with family members who advised they would make final plan-of-care decisions after she had arrived back home.
- The patient continued home health and returned to the hospital for continued aggressive treatment for various organ systems and infections. She admitted to the hospital about every other month and discharged back home on the vent each time.

5.2.13.3   Case Example

- The patient was a fifty-two-year-old Hispanic female who was re-admitting from a nursing home where she had recently been discharged. She had originally been placed in a nursing home due to her vegetative state. From the nursing home, she was admitted to the hospital for cardiopulmonary arrest. She got trach and PEG placement, transferred to an LTAC, where she was successfully weaned, and transferred back to the nursing home. She went back into respiratory distress, back to the hospital, back to a different nursing home. She presented again at the hospital from the new nursing home with her husband, who stated she had developed fevers.
- Ethics was originally consulted during a prior admission where the family had decided to discharge to a different nursing home. Ethics was reconsulted when the patient re-admitted from the new facility. Her husband had a hard time speaking with Ethics again, and he frequently deferred to the children. The children were receptive to hospice this time around and understood the need for de-escalation of care after so many transfers/readmissions.
- The family was distrustful of the hospital's hospice referral after previous experiences at the nursing home. The family dragged their feet over hospice orders, and an accepting hospice was difficult to find due to the patient's vegetative state. Eventually hospice orders were signed and the patient discharged to a different facility.

5.2.13.4   Case Example

- The patient was an ninety-year-old Caucasian female with lengthy past medical history significant for chronic debility with adult failure to thrive. She was bedridden with a PEG tube for nutrition. She had been found with hemoglobin of 4.2 during her routine blood test at the nursing home and sent to the hospital.
- Ethics was called by the nursing home's physician in regard to appropriateness of care for the patient, who was on hospice/palliative at the nursing home.

- Ethics spoke with the patient's family, who explained that the patient had been on hospice but was "discharged when she got better." The family was wary of another hospice referral after the first referral ended in a discharge.
- Palliative and Ethics worked with the family on closing communication gaps and establishing a plan of care. A POLST form was recommended to communicate goals of care to the nursing home to prevent future confusion with nursing home hospice/palliative care.
- The patient was discharged back to the nursing home with a POLST form to prevent future confusing.

End of life can be a confusing time for patients and for family members. Remember that these processes and procedures follow a standard progression for you, but they are new to the family.

1. Know your state laws on payer source allowance for nursing homes, LTACs, and hospice houses. Do not advise what cannot be logistically achieved.
2. I always tell my patients, "If we cannot get a facility acceptance, think outside the box." All procedures, protocols and law cannot compare to extra-human care and effort
3. Know your discharge planners and let them give you a 2 min lowdown on the barriers of discharge for this patient, relative to their technology, caregivers, or payer source. It may not change what you offer, but you will be on the same page.

5.3 Conclusion

You have read all of this and are ready to move on to the next thing, to learn the next procedure or medication distribution—but please heed these words: *If you cannot implement what you have read, at least agree to consult those who can.*

When you suspect brain damage, you consult neurology; when you suspect heart issues, cardiology is called in; and if you suspect needed de-escalation, boundary setting, and better comfort for symptom control, consult the palliative care specialist; those are the clinicians who are trained in the dialogue and discourse of death. If you are too frozen, if you can't communicate, call in someone who can. Knowing that I deal with my patients' and families' constant pain from suffering and death, I am often approached by others who want to share their story. I am not their ethicist or even working their case. Much of the time, their loved one has passed and they need to process and share.

Unfortunately, at the root of their pain is only one gap, one loophole—and that was communication. It was lacking. Someone didn't explain the withdrawal process. Someone didn't say the patient might not make it. Someone didn't use the term "dying." Someone didn't explain what the morphine is for and made them feel like a catalyst and a euthanizer. Someone didn't offer compassion or answer questions. That someone was you. The team couldn't heal the patient, and that's ok—but the providers didn't communicate and that's *not* ok. The common excuse is not having time, inundated with documentation and policy demands—but the literature proves more time spent in one sitting equals less time on recurrent trips into the room: depth in time, not length in time. If you can establish rapport on a bus stop, you can do so as well in the most personal experience of a person's life, the last phase of that life.

References

Back, Anthony, Robert Arnold, Kelly Edward, and James Tulsky. VitalTalk. Available at www. vitaltalk.org.

Beckman, Howard B., Melissa Wendland, Christopher Mooney, Michael S. Krasner, Timothy E. Quill, Anthony L. Suchman, and Ronald M. Epstein. 2012. The Impact of a Program of Mindful Communication on Primary Care Physicians. *Academic Medicine* 87 (6 June): 815–819.

Benton, Kathleen, James Stephens, Robert Vogel, Gerald Ledlow, Richard Ackermann, Carol Babcock, and Georgia McCook. 2015. The Influence of Race on End-of-Life Choices Following a Counselor-Based Palliative Consultation. *American Journal of Hospice & Palliative Medicine* 32 (1): 84–89.

Berman, Amy. 2015. A Nurse with Fatal Breast Cancer Says End-of-Life Discussions Saved her Life. *Washington Post*, September 28, 2015.

Institute of Medicine. 2014. *Dying in America: Improving Quality and Honoring Individual Preference Near the End of Life*. Washington, DC: National Academies Press.

Lakin, Joshua R., Susan D. Block, J. Andrew Bilings, Luca A. Koritsanszky, Rebecca Cunningham, Lisa Wichmann, Doreen Harvey, Jan Lamey, and Rachelle E. Bernacki. 2016. Improving Communication About Serious Illness in Primary Care. *JAMA Internal Medicine* 176 (9, July 11): 1380–1387.

Lowery, Susan E., Sally A. Norton, Jill R. Quinn, and Timothy E. Quill. 2013. Living with Advanced Heart Failure or COPD: Experiences and Goals of Individuals Nearing the End of Life. *Research in Nursing and Health* 36 (4): 349–358.

Neumann, Ann. 2016. *The Good Death: An Exploration of Dying in America*. Boston: Beach Press.

Varney, Sarah. 2015. A Racial Gap in Attitudes toward Hospice Care. *New York Times*, August 21, 2015.

Volandes, Angelo E. 2015. *The Conversation: A Revolutionary Plan for End-of-Life Care*. New York: Bloomsbury.

Epilogue

Because of my long years of experience in living with Daniel and in caring for my patients and their families, this book was in many ways an easy one to write. Those experiences have given me a clear direction and guide for outlining this book.

In other ways, writing this book has been one of the hardest things I have ever done. Every patient in the hospital reminds me in some ways of Daniel, who—contrary to many predictions—was still alive when I began this book, in the most generous sense of that term. He could not move easily from his bed and was totally dependent on those around him—but was hanging on. How? One of his physicians once relayed to me, "Who needs organs!? Not Daniel. His sheer will keeps him here."

We promised him that we support him for as long as he needed us, and we did. As he told us many times, there was something he still needed to do—but what that was, I was not sure. Perhaps it was for my children, for whom he needed to stay here to teach and watch grow up until the final minute. And the Thursday before he passed, they were right with him, watching movies and eating candy while I was away presenting a talk about him. What he refused to acknowledge was the strength he gave to those around him, who watched him suffer with a smile. They did for him because it gave them joy. They built him a room with a place to relax and shower because he amazed them. They listened to his stories and anecdotes, were entertained by his crass humor, and stood in awe of how his spirit stayed intact despite the machines and the narcotics to keep pain almost at bay. Some days, Daniel was angry with me for even saying he was sick, and some days he shared privately with my mother that he knew he was dying. It depended on what he wished to believe in a given moment, and how ok he felt with it. And when he did die, it was just as it should be. He was ordering from Amazon, as he so often did, which left us with treasures to unwrap after his passing. He was in his new comfortable bed, a Sunday with the TV blaring, rotating turns with each parent to watch shows and talk. His respiratory distress was quickly compromised, for absolutely unknown, but legitimately understood reasons, and he asked my parents to call an ambulance. They worked together to do CPR and followed his wishes. He ambulated to the hospital and was put through the resuscitation and shock, just as he wanted. And what did that "futile care" give him? It was not futile; it gave him just what he wanted: last minutes with my parents. They were called back to talk with him and they both got

© The Author(s) 2017
K. Benton, *The Skill of End-of-Life Communication for Clinicians*,
SpringerBriefs in Ethics, DOI 10.1007/978-3-319-60444-2

to tell him they "would be ok" and he could go and enjoy his peace, his eternal party. At that moment, he left with a tear running down his cheek and a smile on his face. At that moment, he taught us that some aggressive care has its purpose, in the patient's phase of death. We must trust our patients.

As much as I hated to watch him suffer through the dialysis and the trach, the vent and the infections, I would never wish for a different end for Daniel. He allowed us to say goodbye and be somewhat ok with the peace he found in death. Through this last phase of a journey, Daniel taught all of us an invaluable lesson, as he always indirectly did: End of life is a *phase* for many patients in this age of technology and advancements. It is not always a moment or an imminent week or day. It may take place over years of a patient's life. That challenges us all the more as professionals to prepare those anticipating grief for what is to come. This means for us that we must offer hope of times together within the reality of the finality of disease that will not weather every storm. No one's time is guaranteed, but the critically chronically ill deserve to understand their outcome and live to the fullest during their end.

If Daniel were someone else's loved one, what would I recommend? I would hope that I would be true to what the patient wanted and needed, but it is hard for family and friends to watch a loved one suffer. I care for the families as much as for the patients, but my ultimate responsibility is to the patient.

As caregivers, we must always keep that in mind.

And what I feel for Daniel now is not the peace that holds me because he no longer suffers. It is this:

> You will not "get over" the loss of a loved one; you will learn to live with it… You will be whole but you will never be the same again. Nor should you be the same. Nor should you want to. —Elisabeth Kübler-Ross

About the Author

Kathleen Benton is the director of clinical ethics and palliative care at a local hospital in Savannah, Georgia. She is heavily involved in the region through volunteer advisory board roles and professorships at Armstrong State University and the Mercer University Medical School. Dr. Benton has a master's degree in medical ethics and a doctorate in public health. She has authored and reviewed many publications relevant to the topics of palliative care, ethics, hospice, and communication. Her first publication, a children's book entitled *Daniel's World: A Book About Children with Disabilities*, is the closest to heart. She lives her vocation, passionate about helping families through ethical decision-making processes. She resides in Savannah with her husband, Rex, and her three children, Julia Grace, Jack, and Andrew.

© The Author(s) 2017   81
K. Benton, *The Skill of End-of-Life Communication for Clinicians*,
SpringerBriefs in Ethics, DOI 10.1007/978-3-319-60444-2

Printed in the United States
By Bookmasters